VOICE OF THE FOREST

神の森

INTERPRETING
THE JAPANESE PERSPECTIVE
ON NATURE

Introduction

"In western culture, you confront the forest and conquer it. In Japanese culture, we think we are embraced by the forest. There is a fundamental difference in the philosophies. No matter how much we progress economically, everyone has some thoughts toward nature and even about a single tree. Within every tree and every forest, there is a *kami* spirit. These beliefs are embedded within Japanese DNA."

Shinto Chief Priest Shigeho Yoshida
of Tsurugaoka Hachimangu

Chinju-no-mori is a sacred type of forest revered since ancient times in Japan. In the forest, various trees suited to the local environment grow in multiple layers in natural harmony. When you travel in Japan, you will sometimes see a thick cluster of trees among the landscape. These are quite frequently *Chinju-no-mori*, indicating the presence of a *jinja* shrine. The *jinja*, always surrounded by *Chinju-no-mori*, plays an important role in forest preservation.

You will absorb quickly how much we love our forests. As much as 70% of Japan's land is covered by the forest, compared to the world average of 30%. Japan is one of the very few "forest nations", and the abundance of forests there can be attributed to the numerous mountains allowing us to keep the high amount of rainfall fostering forest growth. But even more than that, I believe the Japanese recognition of the importance and the role the forest plays is a big part in protecting the forests.

Our ancestors experienced *kami* spirit in the natural world and therefore conveyed their gratitude to the forests, mountains, rivers and oceans. Rituals were conducted at sacred sites in beautiful natural settings where sanctuary buildings (*jinja*) were later erected. *Jinja* shrines are protected and kept meticulously clean, and have a soothing effect on those who visit them.

We have some very notable *Chinju-no-mori* and *jinja*. The sacred forest of Ise-Jingu covers 5,500 hectares, which is approximately the same size as Paris, France. The Omiwa Jinja of Nara prefecture enshrines Mount Miwa at a point where it can restrict visitors from entering. Tadasu-no-mori of Shimogamo Jinja is the *jinja* in Kyoto and is registered as a world heritage. While these are very well-known *Chinju-no-mori*, there are many more sacred divine forests throughout Japan that have been cherished and preserved for generations.

In this book, we will start by explaining the nature of *Chinju-no-mori*. From there, we will interpret the Japanese perspective on nature and how *jinja* are related to forests. We hope you will further gain a remarkable appreciation of Japanese forests through the illustrations and photographs of the *Chinju-no-mori* that exist in Japan today.

Here are some key words to help you understand and think about the forest as we do in Japan.

Chinju-no-mori
鎮守の森

Chinju-no-mori is the sacred forest surrounding *jinja* grounds (compound), in which *kami* dwells. These *Chinju-no-mori* forests have been revered since the beginning of time in Japan. They are seldom logged and many of them have not been cut for hundreds of years. As a result, many trees in *Chinju-no-mori* retain vegetation that is native to the local environment and primeval vegetation can occasionally be found there.

Kami
神について

Reference: Soul of Japan

Since ancient times, Japanese people expressed the divine energy or life-force of the natural world as *kami*.

Kami derived from nature, such as the *kami* of rain, the *kami* of wind, the *kami* of the mountains, the *kami* of the sea, and the *kami* of thunder have a deep relationship with our lives and a profound influence over our activities. Individuals who have made a great contribution to the state or society may also be enshrined and revered as *kami*.

Nature's severity, does not take human comfort and convenience into consideration. The sun, which gives life to all living things, sometimes parches the earth, causing drought and famine. The oceans, where life first appeared, may suddenly rise, sending violent tidal waves onto the land, causing much destruction and grief. The blossom-scented wind, a harbinger of spring, can become a wild storm. Even the smallest animals can bring harm like the mouse that eats our grain and carries disease, and the locust that devastates our crops. It is to the *kami* that the Japanese turn to pacify this sometimes calm but at times raging aspect of nature. Through ceremonies, called 'matsuri', they appease the *kami* and wish for further blessings.

Shinto does not observe a single, omnipotent Creator. Each *kami* plays its own role in the ordering of the world, and when faced with a problem the *kami* gather to discuss the issue in order to solve it. This is mentioned in records from the 8th Century which tell the story of the Divine Age before written history began, and is the basis for Japanese society's emphasis on harmony, and the cooperative utilization of individual strengths.

Jinja
神社について

Reference: Soul of Japan

Japanese people regard the sea, the mountains, the forest, and natural landmarks as places where the *kami* reside. In ancient times, these were regarded as sacred areas, without the need for special buildings, as the *kami* were believed to exist everywhere.

A practice also arose of decorating evergreen trees in sacred courtyards to which the *kami* could be summoned in order to perform rituals. Later, dwellings were built for the *kami* in the forests; structures to be renewed in perpetuity where rituals could be conducted. This is the origin of the shrines known as '*jinja*'. There are more than 80,000 *jinja* in Japan today where various *kami* are enshrined, particularly those who appear in the story of the Divine Age or historical figures known for their great achievements.

Rituals to pray for the peace, security, and prosperity of the nation and community are conducted at *jinja* throughout the year. Prayers may also be dedicated at *jinja* for the well-being of the local parishioners and the guardian *kami* of the community. While these observances are typically handled by members of the Shinto priesthood, individuals will often visit a *jinja* to participate in the yearly cycle of *matsuri*, and on commemorative occasions throughout one's life in order to make wishes and offer prayers of appreciation to the *kami*.

Jinja are sacred places and are always kept clean and pure. Often surrounded by trees, *jinja* are infused with the divine energy of nature. They are places to worship, but also places to relax. Visiting a *jinja*, we feel physically and spiritually rejuvenated. *Jinja* are special spaces for us to reflect on ourselves and express our gratitude to the *kami*.

Contents

- 2 **Introduction**
- 6 **Forests of Jingu**
  神宮の森
- 22 **Chinjyu-no-Mori;**
  鎮守の森
- 30 **Kami of Rice and Trees**
  稲の神、森の神
- 32 **1000 Year Old Sacred Great Ginkyo Tree; Tsurugaoka Hachimangu**
  樹齢1000年といわれたご神木に対する思い
- 42 **Japanese View of The Natural World Reveals Its Logic on Spiritual Sense of Life**
  日本人の自然感から霊的生命感
- 50 **Kami of Water, Mountains, Rice And Wind**
  水、山、稲、風の神
- 52 **Yoshimuta Headwaters**
  吉無田水源
- 54 **Enshrined Water Resourse: Kifune Jinja**
  貴船神社
- 59 **Everything is Alive**
  全ては生きている
- 62 **Sacred Mountains**
  聖なる山
- 70 **A Step at a Time Towards Reconstruction: Yaegaki Jinja**
  再建に向けて一歩づつ！八重垣神社
- 76 **Forests Heal People**
  森は人を癒す
- 80 **Creat Forests For Life**
  いのちの森をつくる
- 85 **The Forest of Meiji Jingu**
  明治神宮の森
- 94 **Summary of Forests in Japan**
  日本の森の現状

Forest of Jingu
神宮の森

Jinja in Deep Forest

深遠な森の中にあるお社

In order to reach Naiku (the central *jinja* of Ise-Jingu), you must cross the Ujibashi Bridge at the entrance of Naiku. The front of Ujibashi Bridge is filled with the aroma of nearby trees and you can feel the energy or "breath" of the forest. This *jinja* where *kami* reside oversees Shimaji-yama Mountain to the left and Kamiji-yama Mountain to the right, and is surrounded by deep, pristine mountains and forests. The forests around Naiku occupy an area of 5,500 hectares, which comprises approximately one fourth of Ise City, an area similar in size to Paris, France.

   Sacred forests comprise ninety three hectares of this area surrounding the *jinja* buildings, and they are forests that have been cherished as a divine zone since the time the *kami* Amaterasu-Omikami was enshrined more than 2000 years ago. Visitors are not allowed to enter the forests, but when they walk through the forests along *sando*, a path built for *jinja* visitors that leads to the main sanctuary, they appreciate the forests' dignified presence.

   This particular *jinja* is widely known as Ise-Jingu, but the official name is Jingu. Jingu is located in Ise City, Mie Prefecture, and is regarded as having the highest status of all *jinja* in Japan. The *kami* enshrined is worshipped as the supreme guardian *kami* of Japan. Naiku is only one of the central *jinja*, but there are 125 *jinja* in total in the compound, collectively referred to as Jingu.
   Besides Naiku (or Kotaijingu), the other central *jinja* is Toyouke-Daijingu (or Geku). Naiku enshrines Amaterasu-omikami, the supreme *kami* of Japan, while Geku enshrines Toyouke-omikami.

Ujibashi Bridge : The bridge is said to connect the human world to the god world. With Isuzu river spanning sixteen kilometers, the bridge is located in the middle at the eight kilometer point. A village is downstream from the bridge.

The torii gate in front of Ujibashi Bridge.

Forest as a Sacred Place

聖地としての森

Jingu originated when Amaterasu-omikami was enshrined more than 2,000 years ago. The treasured and beautiful forest of Jingu spreads out along the slopes of gentle mountain ranges of 300 to 500 meters high. It is estimated that the forest consists of seventy percent evergreen trees and thirty percent deciduous trees. In winter, the forest is dark in color; and conversely, in spring the bright green and orange colors of young oak leaves blend together and give off the appearance of a beautiful carpet. In summer, the leaves turn a deeper green, providing a gentle shade. In autumn, red and orange hues splendidly decorate the mountain slopes.

While this forest and mountain captures the beauty of the Japanese seasons, it is also a valuable forest for its maintenance of a natural ecological system. When you follow the path around the forest, you will feel refreshed and embraced by a gentle feeling. This sacred forest surrounded by mountains is very pure and has an air of solemnity. It is a place where winds from the mountains blow gently and an indescribable divine force can be felt. There is something about Jingu that is different from other forests, including some other *Chinju-no-mori*. What could be so special about Jingu?

It is said in the era of Emperor Suinin, the emperor's princess, Yamato-hime-no-mikoto, received Amaterasu-omikami's oracle. As a result, Kotai Jingu (Naiku) was built.

> Ise Province, where divine winds blow
> a remote but beautiful region where waves
> from distant utopias lap on its shore
> I shall stay in this province forever.

Jingu has been described as "a place that connects modern and ancient times". Rituals have been performed in this forest for two thousand consecutive years. These rituals are done in the same manner as they were the very first time, allowing a practice that was born long ago to live on today.

Throughout further history, Japan went through several wars such as the Genpei War, Warring State Period, Meiji Restoration, World War I and World War II. Despite these wars and their devastating effects on everyday life, Japanese people have offered their prayers of thanks at Jingu. These prayers, offered daily for the past 2,000 years, offer respect, awe and gratitude toward nature and hopeful thoughts to protect the forest. It is the *kami*'s responses to these prayers that fill the forest with divine energy.

The Natural Water Cycle and Its Importance

自然のサイクルの流れとその大切さ

When rain falls in the Jingu forest, the rainwater soaks into the mineral-rich leaf mold. Trees grow big as they absorb water and undergo photosynthesis in sunlight. Various creatures also grow in this forest. Rainwater will slowly go through natural filtration. The filtered water eventually flows down the mountain and becomes Shimaji river, which then merges into Isuzu river. The source of the river lies in Jingu forest.

Jingu property is marked by mountain ridges. The area takes on the shape of a bowl and the rainfall flows down the mountains, passing through Naiku, and to the ocean. It is a pure and sacred stream. This rich forest continually provides water to Isuzu river all year round.

Using water from this river, rice is grown in Shinden (the rice paddies of Jingu), and fruits and vegetables are cultivated in Misono (the sacred fields). Further downstream, salt is made in Mishiohama (the sacred fields for making salt). Finally, the mineral-rich water enters the ocean and nurtures sea creatures.

Utilizing nature, offerings made to the *Kami* are prepared in the traditional manner with an enormous amount of time and effort, pulling water from a well. Fire is made using a *hikirigu* (an ancient instrument to make fire). Salt, vegetable, textiles (silk, hemp), and *jinja* buildings are all made using simple traditional methods that are difficult to imagine in modern times.

With a mountain and a reserved forest, accompanied by a fertile river, the Jingu area sustains itself. If we continue to look at Jingu this way where human beings involve themselves in the natural cycle in an unobtrusive fashion, humans and nature will continue to coexist well.

After visiting Jingu, people often leave with a feeling of spiritual fulfillment. This may be because the way Jingu exists reminds the Japanese people of their origins.

Shinden: rice paddies of Jingu. The photo is provided by Jingu Shicho.

The source of the river lies in the Jingu forest.

Shikinen Sengu

式年遷宮

The Shikinen Sengu is Jingu's largest ceremony. Every 20 years, the two central *jinja*, Naiku and Geku, and all of the 14 *jinja* buildings belonging to Naiku and Geku, the attires of *kami*, the sacred treasures, and the Ujibashi Bridge are renewed.

'Shikinen' means a designated year, and 'Sengu' refers to requesting *kami* to move to a renewed *jinja* building.

This Shikinen Sengu has continued for about 1,300 years. The first one was conducted in the 4th year of Empress Jito's reign (690). It is said that the Empress' husband, Emperor Tenmu, had ordained this ceremony before he passed away. The ceremony did not take place during the Sengoku period (period of warring states) and it was postponed after World War II. Aside from those exceptions, it has been maintained and repeated over the centuries. The most recent Shikinen Sengu took place in Heisei 25 (2013).

When Shikinen Sengu was instituted, some architectural styles had already existed for hundreds or over a thousand years. Horyuji Temple, for example, the world's oldest wooden architectural building located in Nara Prefecture, was originally built in 607 (15th year of the Emperor Suiko's Era) and still retains its original structure after a thousand years. Interestingly, the builders adopted the style of *shinmeizukuri* using *kayabukiyane*, a method of making rooftops made from Japanese nutmeg tree, silver grass, or cogon grass. It is considered to be a primitive roof style that can barely stand tens and hundreds of years of wind and snow.

To address the question of why they used this primitive method, one answer relates to the thought of "*tokowaka*", which translates to "being ever fresh and young". That is, by renewing buildings made of trees with short longevity, *kami* are enabled to always dwell in a place with pristine freshness. Another answer offered is that the method keeps the style cherished by our forebeaers.

No mention has been found in literature as to the origin of the 20-year interval. However, by carrying out the Sengu every 20 years, both the ancient technology and the Japanese tradition have been preserved.

The trees used as lumber for Shikinen Sengu are grown in Jingu in the forest behind the sanctuary. Wood from dismantling the old buildings is reused for the *torii* gates. The Munamochi-bashira of Naiku and Geku are reused for the inner and outer *torii* of Ujibashi Bridge, respectively. Reclaimed lumber is also distributed to *jinja* shrines nationwide for their reuse.

Main building, Shogu of Kotaijingu(Naiku) renewed in 2013.
Photography by Jingu Administration Office.

Misoma-hajime-sai; the ceremony to start the cutting of wood used for the new sanctuary buildings. Photography by Jingu Administration Office.

Kozukuri-hajime-sai; the ceremony to pray for construction safety before it commences. Photography by Jingu Administration Office.

Kanname-sai in October, marching toward the main building at night.

Forest of Trees Used as Lumber for The New Jinja Building at Shikinen Sengu

遷宮のご用材を育てる森

Hinoki ready to be planted.   The ritual performed before planting.   Each Hinoki planted carefully.

Management of forest

Shikinen Sengu requires ten thousand Japanese cypress trees. Builders start with a tree of approximately 40 cm in diameter at chest-height. The tree with the greatest dimensions (at least 122 cm diameter) is used for the sacred door of the main sanctuary. The longest required piece of lumber is 13 m long and is used for Chigi (forked roof finials) in the Yojoden building. Lumber used for Munamochi-bashira (the free-standing columns that support the roof ridge) is the second longest piece at a length of 11 to 11.8 m and 70 to 78 cm in diameter. The trees used for *Munamochi-bashira* take 700 to 800 years to grow in Kiso province. Of all the trees used for the 2013 Shikinen Sengu, about 80% were cut from the national forest in Kiso province. Although Ise is warmer than Kiso, it would still take at least 500 years to grow there. Jingu tries to grow the trees to 1 m (at least 60 cm) diameter at chest-height in 200 years.

Currently in use is a method called "*Juk-oubatsu*" to grow trees with thick trunks for Sengu lumber. With this method, 4,000 trees are planted per 1 hectare of land. In thirty to forty years, differences between each tree becomes evident. The best quality trees are selected and double-marked with paint. The second best quality trees are single-marked.

Per hectare, there are about 50 to 70 double-marked trees and 130 to 150 single-marked trees, with a total of 200 promising trees selected. In order to promote faster growth, trees surrounding the double-marked ones are cut down to allow for more sunlight. Birds or the wind bring seeds of various evergreen and deciduous trees to the forest, and the seeds fall in the cut area where sunlight hits. Those plants will bloom flowers and attract insects and birds, which in turn attract animals. Animal remains become fertilizer as bacteria decompose their waste material. This becomes rich soil, making the mountain very fertile. Tree roots absorb nutrition from the fertile soil. Photosynthesis occurs with sunlight and trees grow thick and rapidly. Biodiversity and the overall ecosystem are carefully taken into consideration in order to create a mixed forest of broad-leaved and coniferous trees.

A forest to pass on to the next generation for another 200 to 500 years

In Taisho 12 (1923), the forest planning project planted *Hinoki* in the Jingu forest, which grew tall enough to cut. The remaining 20% of trees used for the 2013 Shikinen Sengu were cut from the Jingu forest for the first time in 700 years.

During the Nara (710 - 794) and Heian (794 - 1185) periods, lumber used for Shikinen Sengu was cut from Jingu's forest. In the middle of the Kamakura period (1185 - 1333), the number of large trees decreased and trees had to be obtained from neighboring forests. In the middle of the Edo period (1603 - 1868), trees were supplied from the national forest of Kiso (in Nagano and Gifu prefectures).

Moreover, in the Edo period, "Okagemairi" occurred, which were mass pilgrimages to Jingu. At that time, the total population of Japan was 30 million, of which 5 million visited Jingu per year. This Okagemairi happened three times every 60 years during the Edo period. To serve these visitors, the remaining trees in Jingu were cut to be used for firewood or charcoal.

Although they tried to avoid excessively cutting down trees, a large quantity of trees were nonetheless harvested. As a result, the water retaining ability of the Jingu forest severely declined, causing the Isuzu River to flood to 2 m above ground level.

The flood triggered the restoration project in Taisho 11 (1922), and in the following year the forest restoration plan was created to renew the forest. According to this plan, 200 years after planting, the Jingu forest supplies a portion of the trees for use in Shikinen Sengu. 500 years after planting, the Jingu forest will be able to supply 100% of the lumber. It takes 200 to 500 years for a tree to grow large enough to be used as lumber. The trees used for the 2013 Shikinen Sengu were planted by distant ancestors. Their thoughts towards the uncertain future took shape this past year.

Certain measures are taken to ensure trees are not aimlessly cut down. On top of that, prayers of gratitude are offered daily. Combined, the trees selected for use in Shikinen Sengu will be passed on to future generations for 200 to 500 years, along with the traditional methods and attitude.

Here lies the key to humans, nature and *Kami* successfully existing in harmony.

Double-marked trees: A candidate tree to be cut and used as Jingu building materials after two hundred years.

鎮守の森
CHINJU-NO-MORI

Kami reside in *Chinju-no-mori*,
a place that soothes the souls and invigorates Japanese people.

Sadasumi Motegi 茂木貞純

Photo:Kuchira-jinja 朽羅神社 This jinja belongs to Ise-jingu.

CHINJU-NO-MORI is a type of forest where *kami* dwell. To be considered *Chinju-no-mori*, it must be a dense forest at the outskirts of residential area. It must have a forestscape that is dark and heaped up into a mound, and be easily distinguishable as belonging to a *jinja*. At the entrance of the forest is a *torii* gate. Ahead of a long approach to the forest, there is an open space where *jinja* buildings are located. Among the many buildings are *Honden*, the central building where *kami* dwell, and *Haiden*, the building where people worship. In the middle of the approach, there is a *Temizuya* (a fountain at the entrance to *jinja*). In order to get closer to and worship *kami*, one must first be purified, so visitors are able to purify their bodies and souls by washing their hands and rinsing their mouth in the fountain.

When you look around a *Chinju-no-mori* forest, you will notice that the area is very clean and has a sacred aura. The grounds are swept of withered branches or fallen leaves. In a *Chinju-no-mori*, there are large trees, some of which are generations old because it is a forest untouched by human hands for generations. Shimenawa (a braided rice straw rope placed around the objects/trees to indicate the presence of *kami*) is hung around a tree that is markedly bigger than the others. This tree is considered to be a sacred tree to which *kami* hailed down from heaven, and its treetop rises to the sky.

In the ancient times, Japanese enshrined *kami* within untainted nature. They initially worshipped rocks and trees without any *jinja* buildings. These were not ordinary rocks or trees, but Iwakura (rocks to summon the *kami*) and Himorogi (a temporarily erected tree branch to summon the *kami*). Both were considered sacred. In the middle of the sixth century, Buddhism arrived in Japan and temples were built to house statues of Buddha. With this influence, *jinja* buildings started to be built. Although *jinja* buildings were constructed as *kami*'s dwelling places, the Japanese still carry on the belief to this day that *kami* inhabit the forest.

Left photo: *Temizuya* 手水舎 Right photo: *Jinja* building of *Haiden* 拝殿 at Kashihara Jingu（橿原神宮）located in Kashihara, Nara prefecture

"**M**ANYOSHU" is the oldest collection of Japanese poems edited in the middle of the 8th century. It is a compilation of waka, traditional Japanese poetry, written by people of various social backgrounds, from the Emperor to ordinary citizens. In many of these wakas, a *jinja* was referred to as "*Mori*", meaning a forest. Therefore, in these ancient times *Mori* was synonymous with *jinja*. This explains the Japanese understanding of *kami* living in the forest.

Chinju-no-mori are not located in the center of the village, but just outside of it. They are usually located on a small hill overlooking the village, at the edge of the village, or the entrance to the path leading up to a mountain. In other words, they are at the border separating the untouched areas of nature and a civilization. The *Chinju-no-mori*'s locations show us that the ancient Japanese drew a distinct line between the areas where people lived and areas where *kami* lived, with the *jinja* at the borderline. In particular, the mountains were considered to be under the control of *kami*. People were not allowed to enter without a good reason.

It can be said that *Chinju-no-mori* is a place where people meet with *kami* as it is the forest closest to human habitation. In the "Hitachi no kuni fudoki" (a compilation of provincial culture, local myths and histories, edited in the 8th century), there is a folk story about an establishment of a *jinja*. It is a story set near Kasumigaura in Ibaragi-prefecture.

Long ago, a government officer named Matachi was trying to clear reed beds in order to make rice fields. Yato-no-kami came along with many of his fellows, and got in Matachi's way. Yato-no-kami is a snake with a horn on his head and lives in the swamp. It is said that once a person sees Yato-no-kami, the person's family will perish and that Yato-no-kami is a terrifying being. Angry, Matachi armed himself and ran them down to the foot of the mountain and finally killed Yato-no-kami and his fellows. Matachi built a small sacred building and declared "From here up will be the land of *kami*, and below here will be the land for humans to cultivate rice. From now on, I will be the priest of this *jinja* and pray for Yato-no-kami forever. So please don't curse or hold a grudge against me". This was the establishment of a small *jinja*. Many vipers still live near the *jinja*, and people in this area are fearful of snakes and still consider them sacred. The *jinja* enshrining Yato-no-kami still exists as a sacred place of worship.

Small animals that live in *Satoyama* (woodlands located between residential area and the undeveloped nature) are known to the people as the messengers of *kami* because they come and go between the undeveloped wilderness and the village. These include the fox of Inari Jinja, the monkey of Sannousha, the deer of Kasuga Jinja, and the wolf of Mitsuine Jinja and Ontake Jinja.

Right photo: Sacred tree, Ookusu「大楠」It is said that the tree is more than 2000 years old, and is the second largest tree in Japan in terms of the size of the trunk(23.9m). Located in Kinomiya jinja, Atami Shizuoka prefecture.

Sando at Miyazaki Jingu

Dr. Sadasumi Motegi 茂木貞純

Born in 1951 in Kumagaya city, Saitama Prefecture, Japan.
Dr. Motegi received his doctoral degree in Shinto Studies at Kokugakuin University. Dr. Motegi is currently a professor of Shinto culture at Kokugakuin University, Director of the Society of Shinto Studies, a Shinto priest (Guuji) of Komiya *Jinja* in Kumagaya city. In the past, he was head of the Educational Research Center and the chief of the General Affairs Department of *Jinja* Honcho (Association of Shinto Shrines).

Books:
"Shinto to Matsuri no dentou (The tradition of Shinto and Matsuri)" - *Jinja* shinposha
"Sengu wo meguru rekishi (A History of Sengu)" - joint authorship, Shindo Bunka-kai
"Shinji no kiso chishiki (Basics of Shinto Rituals)" - joint authorship, Kodansha Ltd.
"Nihongo to Shinto (Japanese Language and Shinto)" Kodansha Ltd.

Throughout the year, several rituals are performed at the *jinja*s. Although each *jinja*'s ritual is original and follows its own tradition, the rituals are performed based on the cycle of rice cultivation because the result of the rice harvest affects the prosperity of the community. Therefore, worshippers all pray in the spring for a bountiful harvest for the year, harvest the crops in autumn, and then give thanks for a good harvest in the winter.

Following are the annual events for rice cultivation in a typical farming village:

In the early spring, rice farming begins with plowing the field. This process involves cultivating the soil, digging the field very deep, mixing the dirt with fertilizer, and allowing the dirt to infuse oxygen. Around this time, all the village people help each other to clean and maintain the man-made water channels used to supply water to the fields during rice planting.

When spring is well underway, they sow the seeds in a nursery in a small bed until they are 15 cm high, and then transplant the seeds into the rice fields. During this time, people hold spring rituals praying for a bountiful harvest. Everyone gathers at the *jinja* with offerings and *sake* (rice wine) and pray to *kami*. After the ritual, offerings and *sake* are withdrawn and shared among everybody as a way to celebrate. At this time, they also discuss important topics and make decisions for the community. By discussing and making decisions in front of *kami*, they have a special meaning.

After one month in the nursery, the rice seedlings grow to 15 cm and are transplanted into the rice field in May. *Ofuda* (talismans) from the *jinja* are placed at the spout of water channels. Rice cultivation is something that *kami* taught humans, so there are some areas throughout Japan where rice planting itself is performed as a ritual. The rice fields are then watered and plowed for planting. Young girls dress up in folk costumes and plant young rice sprouts in the prepared field while music plays in the background. Everyone in the community participates in this event.

Young rice plants basking in the early summer sunshine grow quickly and strongly. When the rainy season starts and summer begins, farmers must manage the water and remove weeds. If they are blessed with good weather and there is no damage from a lack of water, cold weather, typhoons, or harmful insects, flowers will bloom from the rice plant and the grains will appear at the peak of summer. One month later, as fall approaches, they will reach harvest time. By October, all rice harvesting in Japan will be complete.

Farming processes are concluded in November to prepare for winter. A ritual is held to show appreciation of the harvest. This ritual is called Niinamesai or Shimotsukisai. People purify their bodies and minds and show their gratitude towards *kami* by preparing new crop rice, *sake*, and delicacies from the land and sea. Folk entertainment such as the Kagura dance and the Lion dance are performed for *kami*. Ritual music is played all night long and people communicate with *kami*.

Japanese sense *kami* in every aspect of nature and therefore enshrine *kami* in nature. Yao-Yorozu-no-kamigami (a myriad of *kami*) are still present in Japanese myth. Among those *kami*, the highest and most noble *kami* is Amaterasu-omikami, symbolized by the sun. The *kami* from the myths are enshrined in various *jinja* throughout Japan.

Since the Yayoi period, Japan has developed, with rice farming as our main industry. Therefore, we work hard with nature and manage to survive. The custom of working with others is closely related to rice farming. Working hard is one of the teachings of *kami*, in that it leads to a good life and without it, there is no other happiness. Still, no matter how hard one works, a single typhoon can cut the harvest in half. There are many kinds of disasters that hit Japan. Earthquakes, volcanic eruptions, and tsunamis can hit once every several years. So while nature provides the harvest, at the same time nature can be terrifying. Therefore, Japanese people know there is no other way but to coexist with nature, neither intending to manage or control it.

Since the modernization of the Meiji Restoration, the way of thinking about nature changed for a time. The economic value of rich wood resources of *Chinju-no-mori* became a priority, thus many aged trees were logged and *jinja* lands were sold. It was disastrous for the forests. However, people gave increasing attention to environmental issue and they again became interested in the wisdom of their ancestors. I believe there is still something we can and will learn from our history of creating, raising and protecting the *Chinju-no-mori*, for it soothes the souls and invigorates Japanese people.

Rice field before sunset in Niigata Prefecture.

Amaterasu-Omikami Awarded Rice to The Nation

稲を授けた神　アマテラス
日本神話、稲にまつわるお話

According to Japanese mythology, when sending the *kami* Ninigi-no-mikoto to this world, the *kami* Amaterasu-omikami sent him with three sacred treasures and with rice to feed the nation.

Based on this mythology, the rice now made throughout Japan is gifted by Amaterasu-omikami. This may explain why the Japanese consider rice farming so important. Rice farming requires water; it is impossible for rice to grow where no water flows. Therefore, since ancient times Japanese have protected and conserved water resources and created forests. The Japanese people have planted trees in the mountains for several generations due to the fact that, as a volcanic island, Japan has soil that is easily washed away. It takes tens to hundreds of years to successfully build a forest. Ancestors patiently planted forests for the generation after them and for the country, even if they received no benefit during their time of a fully developed forest.

The forests create rivers with a large amount of water; fish thrive in the ocean downstream, enriching people's lives. In other words, it can be argued that rice planting creates forests, which in turn shape our culture. Amaterasu who taught rice-planting can be said to be the *kami* of rice.

From The Nihon Shoki (Chronicles of Japan), Yoro 4 (720AD).
Under the supervision of Sadasumi Motegi 茂木貞純
Illustrated by Mihoko Inazawa 稲澤美穂子

Kami of Tree, Susanoo and His Son Itakeru
木の神スサノヲとその御子神イタケル

Itakeru

The *kami* Susanoo had a problem with his roughness in Takamagahara (heaven) and was punished by being banished from Takamagahara and sent to the province of Izumo, Japan. In Izumo, he slayed a big snake and obtained the Kusanagi sword from the snake and offered it to Amaterasu-omikami. This story is one of the most famous Japanese myths, well known as "Yamata no orochi".

When Susanoo descended from heaven to Izumo province, he was with his son, Itakeru. Itakeru brought many types of trees and planted them from Kyushu throughout Japan to make the entire nation green. Itakeru was praised for his work and was enshrined at Itakiso *Jinja* in the Kii area of Wakayama prefecture.

Susanoo's tree planting

Excerpted from the NihonShoki, modern language version:

Susanoo said "there are rich treasures of gold and silver in Korea. In my country's posterity, we must have boats as our floating treasure." He pulled out his beard hairs and scattered the hairs. The beard hairs immediately turned into cedar trees. Then he pulled out his chest hairs and scattered them. The chest hairs turned into cypress trees. His buttock hairs turned into podocarpus (Japanese Yew) and his eyebrow hairs turned into camphor trees. Susanoo indicated directions for the use of each tree as follows: cedar and camphor for good quality boats, cypress for the materials of a palace, and podocarpus for coffins.

This story of Susanoo pulling his body hair out to plant the trees needed for Japanese people, showed them how to use the trees. Research proves that some parts of old coffins were made by podocarpus, as directed by Susanoo. Even in the present day, cypress is used for *jinja* buildings such as at Ise-jingu or other *jinja*s in Japan.

Tsurugaoka Hachimangu
鶴岡八幡宮

1000 Year Old Sacred Great Ginkyo Tree

樹齢1000年といわれたご神木に対する思い

"The Sacred Great Ginkgo Tree at Tsurugaoka Hachimangu was uprooted in March 2010."

The news traveled around not only Japan, but also the world. People were surprised by the news of the tree's uprooting, and many visited to mourn its loss. Even foreign media became interested in the news. Why did the fall of a tree have such a large impact on the Japanese people?

We interviewed Shigeho Yoshida, the chief priest of Tsurugaoka Hachimangu.

Profile of Chief Priest Shigeho Yoshida 吉田茂穂

Born in Kobe City, Japan, on April 15, 1942
Graduated from Kokugakuin University, BA Shinto Study, March 1966
Currently Chief Priest of Tsurugaoka Hachimangu since April 1997
Currently Managing Director of the Association of Shinto Shrines since June 2010

Activity
2000 Performed Kagura in Strasbourg, France.
 Delivered a speech on Shinto philosophy at a symposium promoting peace among religions and races
2001 Performed in Yabusame ritual ceremony in London
 Served as a guide to Prince Charles from the UK and Prince Naruhito from Japan
 Delivered a lecture at City University London
2006 Visited the Vatican as a member of the Kanagawa Religious Organization
 Met with the Pope
2006 Performed Kagura for Priest Chogen's 800 year memorial ceremony in Todaiji Temple in Japan
2009 Delivered a lecture on Shinto in Washington, DC
2009 Delivered a lecture on Shinto at the University of Maryland
2009 Delivered a lecture on Shinto at the George Washington University
2010 Delivered a lecture on Shinto at Beijing Foreign Studies University
2013 Delivered a lecture and performed Kagura at the University of Hawaii at Hilo

Priest Shigeho Yoshida (referred to as Yoshida): "It was March 10, 2010, the cold time of the year. That day, there had been continuous rain and sleet from the night before. I was at my house on Mount Daijin behind Tsurugaoka Hachimangu. Around 5 a.m., before dawn, a staff member called me up and exclaimed 'Priest Yoshida, a terrible thing has happened. The great gingko collapsed.'

When I skeptically went to the stone steps in front of the main sanctuary, I could hardly believe my eyes. I tried to walk down the stone steps but couldn't do it well because my knees were shaking. I somehow managed to walk down. At the site, other staff members who had also hurried over seemed awestruck by the scene.

The first thing I said was: 'Prepare a heavy equipment right away'. I don't actually remember saying that, but I assume I said that because I didn't want to leave the tree laying there to be seen by visitors. Of course, it's impossible to prepare that kind of machinery right away. One of the staff members brought some white sheets and used it to create a wall around the tree to hide it from people's eyes, though the tree was still completely visible once you went up to the top of the stone steps. After that, I spent the rest of the day rubbing the tree with thoughts of gratitude toward it."

Interviewer: "You were rubbing the tree all day long?"

Yoshida: "That's right. Meanwhile Dr. Hamano of Tokyo Agricultural University came. He asked me how I wanted to dispose of the tree. He said 'The root is completely dried. There is no sense in trying to bring it back to life so it is better to deal with it soon.' I thought that if it can't be saved, I need to think of what to do with the tree.

Right then, one of the visitors came to me and said, 'This is so terrible, priest, please you must try and bring this sacred tree back to life.' Before hearing these words from the visitor, I was depressed. Then I had the feeling of gratitude toward the Great Gingko. I was very thankful that the Great Ginkgo was there. To be given such positive words by a visitor at a time like that gave me relief, and my outlook changed. Hearing his plea, I changed my mind and decided to try to save the tree if possible, even if it might not work.

I told Dr. Hamano to give me a little time to decide, and told our staff to prepare the gratitude ceremony for the tree. *Saigi-ka*, the section in charge of rituals, was asked to do a ritual of appreciation. I thought we had to offer prayers for the Great Ginkgo to thank it for protecting us for thousands of years.

Late afternoon of the day when I started the ritual alone, I felt something odd. I felt that there was someone else behind me and I turned around to see all our staff members lined up joining the ritual. I had not asked anyone for their attendance. As a priest of Tsurugaoka Hachimangu, I just wanted to give my gratitude to the Great Ginkgo for protecting our *jinja* and intended to do so myself. I then realized that, as it turned out, everyone

else had the same feeling.

Before I knew it, a helicopter was flying over our *jinja*. It was NHK (Japan National Broadcasting) filming coverage about our tree. The coverage was broadcast almost immediately. White-robed *shinshoku* (priests) and maiden's red hakama* projected their silhouettes in the evening mist and it looked like a beautiful art piece.
*hakama: loose trousers forming part of Japanese formal dress.

Many viewers of the TV broadcast called NHK and us, asking if Tsurugaoka Hachimangu does all of that just for a tree. We also received phone calls from Honolulu in Hawaii and Tel Aviv in Israel. The news seemed to travel around the world in a blink of an eye. It was televised abroad, but since Japanese viewers who were abroad didn't comprehend what was said in the news in the local languages, they called us to inquire about it."

Interviewer: "I assume many Kamakura citizens were shocked, weren't they?"

Yoshida: "Quite a number of citizens were shocked. The tree's existence was a part of their lives. When I spoke with them, as they tried to hold back tears, many of them requested that the tree be regenerated. There were others who asked for a small piece of the tree for themselves.

During the new year holiday, the weather was very nice. Towards the end of January, it began to get very cold and the number of visitors to Kamakura city decreased. In Febru-

Right photos from top: Ornament for the Star Festival. Middle: jinja's maidens. Bottom: Shirahata Shrine 白旗神社 Enshrined Minamoto Yoritomo and Minamoto Sanetomo. Both of them were shougun in feudal Japan.

ary, the quiet days continued. After the great ginkgo fell, plenty of people came unexpectedly. Then the people of Kamakura city told me that they were thankful for the great ginkgo bringing many people. At that time, I was sad, lonely and having such a hard time, that I was upset at the thought of being thankful for the tree being fallen. [As he said this, Yoshida laughed.]

Anyhow, after the tree fell, there were more visitors than there were during regular years.

Later we began to put a registration book for visitors to write their name, address and what they wished for the tree. Within a month, a total of 60,000 people had signed it."

Interviewer: "What was the condition of the Great Ginkgo Tree when it fell down?"

Yoshida: "Around this area, we always get wind blowing from the south. This tree grew up leaning against the wind. The tree was leaning south and against the wind for balance. An unusually strong wind from the north caused the tree to fall down. Normally, this tree should have fallen southward because of inclination. But for some reason, the tree instead fell westward by twisting its body as if to avoid falling southward which would have demolished the roof of Maiden, lower worship hall. As a result, Maiden was not damaged and with a time of collapse at 4:46 a.m., no one was injured."

Interviewer: "The way it fell looks like it was arranged by *kami*."

Yoshida: "There were some interesting facts revealed after the collapse of the tree. The tree was 30 meters high, with a girth of 8 meters. The tree was hollow up to a height of about 25 - 26 m. Five rod-like objects ran through the hollow area. These were called "*Jine*" (scion root). One of the *Jine* was soft and spongy. The rest of them were like logs of a cedar tree. These *Jine* were snapped by the shock from falling. The *Jine* worked like the umbilical cord of a baby. The tree absorbed nutrition through the *Jine* from the ground and distributed it to the sprout of the young trees growing every year. The tree tried hard to preserve its seeds.

However we were not aware of this. We kept cutting the sprouts every year. We were told to cut them because the sprouts will use up the nutrients of the parent tree.

One more fact we learned was that the tree had more than one set of growth rings. Normally a tree has only one set of growth rings, but the Great Ginkgo had extra sets of small rings around the main set of rings. This was probably because sometime in the Edo period they allowed a budding young tree to grow on the Great Ginkgo and they both grew together as if the mother tree was holding its baby within her trunk."

Interviewer: "It's as if we are witnessing the preciousness of life itself?"

Yoshida: "Trees hardly die standing when their life ends. They fall down and disintegrate into the soil to become the fertilizer for the next generation. When I realized that the Great Ginkgo had tried to preserve its seeds, the

Left photo: Tsurugaoka Hachimangu;
It is said that 1.8 million people visit per year.

fallen tree looked so divine and solemn. It is hard to describe it with words but it was awe-inspiring and caused me to shed tears. We tend to cut down trees because they are in the way, but I think every tree is a life form that has its own energy to continue living. A forest is a group of these living trees. So when I think about it, I learned the importance of preserving each tree."

Interviewer: "I heard there were people from other countries wanting to cover the story."

Yoshida: "After a while, the British magazine Economist came for an interview. Since their magazine is about economics, I asked them why they had come. They said 'The news video was broadcast in London and became a topic people wondered about. Japan is such an economically developed nation. Its manufacturing industry has received high praise worldwide. We don't understand why such an advanced nation would go through all this trouble just for a tree. That is what brought us here to inquire.'

I explained the basic philosophy of the Japanese to them. I said, 'In western culture, you confront the forest and conquer it. In Japanese culture, we think we are embraced by the forest. There is a fundamental difference in the philosophies. No matter how much we progress economically, everyone has some thoughts toward nature and even about a single tree. Within every tree and every forest, there is a *kami* spirit. These beliefs are embedded within Japanese DNA.' They seemed to understand."

Interviewer: "Were there any others from foreign countries?"

Yoshida: "There was a visitor from the United States who came to see the fallen great ginkgo. He looked disappointed. I asked him what he thought of it, and he responded by asking me, 'Do people really have special feelings about this?'

He was from California and said that there was a much bigger tree, called the 'giant sequoia' in Yosemite park. The giant sequoia is the world's largest tree. Compared to that, this ginkgo is small.

I explained to him that it is true that the size is much smaller than the giant sequoia, but the Japanese strongly believe that the spirit of *kami* dwells within this sacred great ginkgo. This ginkgo tree has watched over Japan's history. I also talked about the relationship between trees and Japanese. He said he understood and left."

Interviewer: "What do you mean by the relationship between trees and Japanese?"

Yoshida: "I talked to him about Japanese myth. Izanagi-no-Mikoto and Izanami-no-Mikoto were two of the primary *kami* in ancient Japan. The islands of Japan were formed, Japanese ancestors were born, and mountains, rivers, plants, and all creatures were created. All of the creatures came from the same *kami*. According to the view based on this mythology, trees, human and *kami* are all brothers and sisters. Western culture may stand on the history of conflict to create your own future. We Japanese have the philosophy that 'we live in harmony with nature' so we treasure even a single tree."

Interviewer: "Currently the part of the original tree is placed next to the young tree, isn't it?"

Yoshida: "We cut a 3.7 m long piece of the original tree from the root and placed it beside the young sprouts. Of the sprouts, one is growing as a tree very steadily and well now. The young sprout is not stable yet.

    We offered the prayers for regeneration on April 10. It's not like I asked anybody to come. I was only planning to do it privately in the *jinja*. Ultimately, there were over a thousand people who attended the prayer.
    Many thoughts passed through my mind during the prayer. I was impressed with the solemnity and dignity of the great nature. Under this Great Gingko the historically famous assassination of Shogun Minamoto no Sanetomo* had taken place. I wondered what kinds of scenes this Great Ginkgo saw throughout its near-1000 years of history. The tree must have witnessed samurai coming and going between the Kamakura and Edo period. Saigyo** may have looked up at this tree. Both before and during World War II, there were many soldiers that visited this *jinja*. Many soldiers departed to war from here. The Great Ginkgo was a witness to all of this history.

*Minamoto no Sanetomo - Third Shogun of the Kamakura Shogunate (1192-1219)
**Saigyo: Famous Japanese poet (1118 -1190)

    While I was talking about how this tree had witnessed such history, I felt overcome with emotion that I couldn't even speak just then.

    Moreover, I came to realize just how everyone is so supportive. Because of their encouragement and support, I have come to believe everything will be fine."

Young sprouts.

Original tree is placed beside young sprouts.

Priests marching for a ceremony.

The ritual of Yabusame.

Kagura dancing dedicated to kami.

The ceremony of "releasing living creatures".

Tsurugaoka Hachimangu 鶴岡八幡宮
2-1-31 Yukinoshita Kamakura Kanagawa-ken

Tsurugaoka Hachimangu was originally built in Kohei 6 (1063) to enshrine Hachiman kami as a tutelary deity of the Minamoto clan in Zaimokuza, Kamakura city. Shogun Minamoto no Yoritomo relocated the jinja to its current location in Jisho 4 (1180) and built Tsurugaoka Wakamiya.

The ceremony of "releasing living creatures" is one of the greatest rituals of Hachiman kami and is strongly influenced by Buddhism. Hotaru-hojo-sai (Releasing of Fireflies) and Suzumushi-hojo-sai (Releasing of Bell Crickets) ceremonies are held in June and September, respectively. These kind of rituals were discontinued by the Ordinance Distinguishing Shinto and Buddhism in Meiji 1 (1868) by the Meiji Revolution Government and restarted again in Heisei 16 (2004).

Untouched nature must be there for the fireflies and the bell crickets to survive. Hence, clean and beautiful nature must exist for these ceremonies to be performed, as these ceremonies were restarted in part to help people realize the importance of preserving nature.

Tsurugaoka Hachimangu invites ambassadors of foreign nations and international students to the rituals to actively promote Japanese culture.

Another well known ritual is Yabusame, a mounted archery ceremony. In 2001, the Yabusame ritual ceremony was performed in the United Kingdom at London Hyde Park and was observed by Prince Naruhito and Prince Charles.

The Tsurugaoka Hachimangu worshippers' association, named "Enju no kai", has a unique activity. They hold activities and events for the community that teach them Japanese culture and promote international exchange as well as events catered for children.

Tadasu-no-Mori
糺の森

Tadasu-no-mori at Shimogamo Jinja has been famous for its unique history. It is located in north of Kyoto where Kamo and Takano river meet. The forests have maintained the natural vegetation since its foundation. It is said it has more than 2100 years of history.

Tadasu-no-mori is a National Historic site, a Natural Heritage site, and a U.N. World Cultural Heritage site of its own right.

Shimogamo Jinja（下鴨神社）
located at Sakyo-ku, Kyoto

"Chinju-no-mori"
Kami reside in untainted natural world
無垢の自然に神が宿る、「鎮守の森」

Japanese view of the natural world reveals its logic on spiritual sense of life

日本人の自然観に見る霊的生命感

An essay by Minoru Sonoda 薗田 稔

" The forest's bounty is receiving of life and it's the life of *kami* you are given."

To learn about "*Chinju-no-mori*" (sacred forest attached to *jinja* building) is to learn about the root of our Japanese soul. Learning will lead you to a grand theme that questions mutual survival of humanity and nature which encompasses environmental issues. We interviewed Mr. Minoru Sonoda, who is well known for his work in restoring and regenerating forests and is active in education of "*Chinju-no-mori*" through the non-profit organization Shasou Gakkai (Society of Forest in *Jinja*). He is also a Professor Emeritus at Kyoto University and the Head Priest at Chichibu Jinja.

Water Cycle
水の循環

Japanese have been treasuring the forest since ancient times. Communities were formed based on the forest, using water from the forest to cultivate fields. They believe *kami* dwell in these rich forests. Living in harmony with the forest is the way of life of Shinto. In other words, the forest can be called "mother of water and life".

The water flow starts from the secluded mountain, then to the mountain near the habitations, then through the habitations, onto the plain, and finally to the ocean. The water turns into vapor which makes the clouds. The clouds then make rain and water falls back onto the earth. The forest is the foundation of the great water cycle. Japan is a mountainous archipelago and these mountains serve as an important source of water.

During the Jyomon period (14,000 BC - 300 BC), people basically lived in the forest. Rice farming then became the basis of life in the Yayoi period (300 BC - AD 250), requiring irrigation and a sufficient water source for the rice paddy fields. In order to have ample water downstream for farming, people needed a forest upstream as the water source, and to cultivate it. As a result, people have taken good care of the forest upstream in order to live in the village without worrying about being deprived of water.

To Enshrine Kami
神を祀る

Kami never appear in front of people. It is not idol worship. It is believed that *kami* reside in chaste and unscathed nature such as rich virgin forest, lively trees, power-emanating stones; all these untainted natural matters with abundance of life.

Kami actually reside upstream of the community, in the remote mountains. However, Japanese people built *jinja* near the villages where people lived and enshrined *kami* there. When you visit *jinja*, you won't see anything. The doors of the main sanctuary are usually closed. In front of the door, a *gohei* (ritual wand) and a mirror are placed but no figures or shapes of *kami* are present whatsoever. Visitors from outside Japan are surprised at this fact, but it is as expected to the Japanese people. When you visit *jinja*, you may expect to see a magnificent statue of *kami*. In Japan, the most sacred matter will not show their figures, but you can nonetheless feel its being. It is a characteristic and one of the essential qualities of sacred matters in Japan. It cannot be detected by human's five senses, yet there is something venerable. Despite the influence of Buddhism or Taoism, the sense that *kami* dwell in nature still remains in people's minds to this day.

Chichibu Jinja was established in the 11th year of reign of the tenth emperor Sujin (estimated to be around 2,100 years ago). Since then Chichibu Jinja stands as the head *jinja* of Chichibu region. Chichibu night matsuri festival has 350 years of history dating back to the *Edo*-era. It is a well-known grand event which draws over 200,000 visitors. Chichibu-city gets illuminated by fireworks and parading floats with lantern decorations.

JINJA BUILDINGS AND FOREST
社殿と森

The Kanji character " 杜 (*jinja's* forest)" is comprised of " 木 "(tree) and " 土 " (dirt). " 杜 " represents a bountiful forest where soil is fertilized and water accumulated by an abundance of trees. The forest itself is considered sacred so when one pays a visit to worship *kami* there, they offer *sakaki* (a sacred evergreen tree of which usually branches are used) and hang *shimenawa* (a braided rice rope) to show that the site is in suitable shape for *kami* to reside. This means that there are no *jinja* buildings there. In other words, it is the forest that makes a *jinja*. Without a forest, there is no *jinja*. Therefore, *jinja* in the village must be surrounded by a forest called "*chinju-no-mori*".

When *matsuri* ceremonies are celebrated at a *jinja*, a ritual of inviting *kami* is at the beginning of the worship and then a sending-off at the end. During these worship ceremonies, they use sacred object such as branches of *sakaki* or the *gohei* for *kami* to reside in.

It was not until the Asuka Period (593-645) that the *jinja* areas began to erect buildings. This was especially influenced by the influx of Buddhism from China. In addition, many powerful clans started to live in buildings with raised floors around this time. Therefore, erecting *jinja* buildings became a norm. Fundamentally, it is still the forest which makes a *jinja*. Today, *jinja* buildings seem to get all the attention of the public as a place of worship, yet the presence or lack of forest in a *jinja*

Dr. Minoru Sonoda 薗田 稔
Chief Priest of Chichibu Jinja
Professor Emeritus of Kyoto University
Vice Chairman of Board of Directors of Sha-sou Society
Books:
"Matsuri no genshogaku
(Phenomenology of Shinto)"
"Daredemo no Shinto
(Shinto for everyone)"
"Shinto"
"Bunka to shiteno Shinto
(Shinto as a culture)"
Kobundo Publishing

should be the more important matter, as it is the sacred land.

Chinju-no-mori allows one to be in touch with nature. One can also be in touch with the unseen, however indescribable it may be. These are important characteristics of Shinto and make up the primary difference between Shinto and all other religions. *Chinju-no-mori* and *jinja* are what connect human with nature.

Forest of Natural Vegetation
自然植生

Chinju-no-mori is where you will see most native trees remaining. Other lands tend to plant and grow trees that are more economically oriented, like farming. Since a *jinja* does not expect to gain economically from its forest, it tends to have vegetation more suitable and domestic to its location remaining intact. From low lying grass to small trees, medium height trees and tall trees, self-sustaining balanced vegetation usually is still present in most *chinju-no-mori* in *jinja*. Of course there are some *chinju-no-mori* that may have been created after economically utilized land was restored and its forest regenerated. This means that not all *chinju-no-mori* is natural forest. It is not easy to keep everything completely natural.

The main theme of the academic book written by Dr. Akira Miyawaki is "To Make *Chinju-no-mori* Into The Forest of the World." In reality, everywhere Dr. Miyawaki visits for his tree-planting program, the first site he visits in each location is the *chinju-no-mori*. He tries to find what are the most suitable trees and vegeta-

Walking the approach

When walking to the *jinja* building, it is best to have an approach that is surrounded by thick forest and has the shade of trees. While walking along the approach, you can sense your feelings change as you relax and release yourself from daily troubles. At the temizuya (water fountain at the *jinja* entrance), you wash your hands, rinse your mouth, and finally enter the *jinja* sanctuary. A *jinja* approach can provide richness to a visitor.

tion for planting by examining the natural vegetation and the potential natural vegetation in these *jinja* forest.

He takes seeds from these *chinju-no-mori*, plants the seeds in a tray to grow into seedlings, transplants the seedlings into a pot to grow into saplings, then transplants the saplings to the tree-planting site. By following this method, trees grow without special care and flourish because they are already suited to the local climate. If trees are brought from other locations, they need special attention. Locally suited trees are already pre-screened and the toughest.

Role and Necessity of Chinju-no-mori
鎮守の森の役割と必要性

There was a big fire in Chichibu city on March 20, 1878 (Meiji 11). During the fire, trees surrounding Chichibu Jinja protected the *jinja* buildings from the ravages of the fire. After the fire was subdued, the forest served as an evacuation area. The forest can take on a disaster prevention role rather than being just open space. Making the forest was wisdom carried forward from ancient times.

Another importance of the forest is its role in fertilization of soil and securing groundwater. Research conducted by Tokyo Metropolitan Government reported that *chinju-no-mori* plays an important role in cleaning the air. *Chinju-no-mori* actually helps prevent the urban heat island effect. It contributes to absorbing carbon dioxide and giving off oxygen. Most city parks do not help significantly, but *chinju-no-mori* has been proven to have these effects.

Destructions of Chinju-no-mori
鎮守の森の破壊

Most *chinju-no-mori* have records of being well preserved until the Edo period (1603 - 1868). Since the Meiji Restoration however, the Japanese sense of reverence toward these forests has unfortunately waned. As a result, records show destruction of many *chinju-no-mori*.

Our *chinju-no-mori* at Chichibu Jinja, named "Hahasono-mori," has been beloved by people. It was once a lot more lush, as Taisho period (1912-1926) photographs show the forest was quite dark even in daylight at that time.

However, when the local government decided to widen the road, they forced the trimming of our forest instead of those at houses opposite the street, possibly to ease negotiations. As such, when there were improvement plans to allow roads and train tracks to go through *jinja* forests, the head priest gave his consent. A number of forests were lost because of such city planning. In retrospect, this was a big loss.

While there are records of these lost *jinja* forests in the cities, there are also numerous records of creating forests. Throughout Japan, several forestation and irrigation efforts have been carried out. This was especially the case in areas where the lack of forests nearby created an apparent water shortage. For example, in Yoshimuta Highland in the Aso Mountain range, several generations of forestation effort were carried out over tens and hundreds of years to help create irrigation canals for the rice fields. Generations of local communities planted trees in hopes of developing a forest sufficient in supplying water for irrigation canals. They built a small *jinja* building in the forest where they planted these trees within which *kami* of the mountain and *kami* of the water were enshrined, and Matsuri ceremonies and festivals were held to commemorate these *kami*.

SPIRITUAL SENSE OF LIFE
霊的な生命感

Japanese people treasure the forest and trees. On top of that, we believe that human beings rely on other life forms to survive. We take the lives of other beings in order to sustain ourselves. The Japanese way of thinking is that everything in this world has some form of life and a soul.

In Japan, we build memorial monuments and/or small *jinja* to hold ceremonies to commemorate everything to which we feel we owe our gratitude. We hold memorial services annually for the departed souls of donated human cadavers and experimental animals at university medical and pharmaceutical colleges and laboratories.

We even have memorials for the trees and grass as well as memorials for insects. Memorial towers for trees exist in Yonezawa City in Yamagata Prefecture for the vast amount of lumber that was logged to rebuild Edo city (currently Tokyo city) after the Great Edo Fire. Several memorials still remain along the rivers in Yonezawa city.

A memorial tower for insects exists along the Tadami river in Nishi-aizu. In this area, insects in the rice fields are exterminated. Since the Japanese believe that even small insects are precious lives, they pray for the killed insects, hence the memorial tower. The memorial tower was designated an intangible cultural asset for the Fukushima Prefecture.

In the Koya Mountain cemetery, there is a memorial tower for termites. It was built by a termite exterminator company. Other well-known memorials include fish memorials in fishing ports and memorial services held on midsummer days for *unagi* (eel), which are a famous delicacy of the Japanese. Needle and doll memorials are also common folk events all over Japan. These memorials are testament to the fact that the Japanese believe they are kept alive by sacrificing these other lives, and therefore, owe thanks to every form of matter. As seen from the custom of having these memorials, it can be said that Japanese believe that we as human beings should not live our lives selfishly.

Material Civilization to Life Civilization
物質文明から生命文明へ

The Japanese take good care of the forest. It is not only because *kami* reside in the forest, but also because they believe that many lives exist and spirits dwell in every part of nature. Such a spiritual sense of life subsists in Japanese people.

Modern civilization is considered a materialistic civilization. We keep saying that we need to change our direction and work toward a spiritual civilization. However, it is difficult to refrain from a materialistic lifestyle. As an old Japanese saying goes, "A samurai pretends he has eaten well, when in reality he has no food to eat". With a shift to a spiritual civilization, it is unrealistic to completely deny everything materialistic. It is important for some materialistic affluence to be allowed while we also fulfill ourselves spiritually. In this regard, it is important that we value lives and make life more fulfilling, realizing that we as human beings are kept alive by consuming and using other forms of life, whether they be animals or vegetation. Thus we should treasure all living forms on this earth, not just human lives.

I advocate that we should move on from being a materialistic civilization to a life civilization. In modern terms, we should build ecological affluence. It would be good if we can think in terms of a civilization where we can keep every being alive.

Aerial view of *jinja* in the Yamato basin of Nara prefecture
奈良県、大和盆地　立体的な祭祀の世界

Kami of Water
Kami of Mountains
Kami of Rice Fields
Kami of Wind

Under the supervision of Minoru Sonoda 薗田　稔
Illustrated by Mihoko Inazawa 稲澤美穂子

←Naniw

Emperor Jimmu, the first emperor of Japan, established the first imperial dynasty of Japan in the Yamato basin. His enthronement is said to have taken place around 660 BC.

Between the 4th and 7th century, powerful clans or kings cultivated this area over several generations. During cultivation, it was important to control flooding and irrigate the wasteland. Therefore the powerful clans or kings enshrined *kami* at the key points of the nature world: mountains, valleys, rivers or fields.

These are the *kami* of mountains, wind, water and fields and have been enshrined to this day.

The Yamato dynasty greeted all of these *kami* and conducted rituals at the imperial court.

Similarly, Hirose Ooimi Jinja greeted these *kami* as well as the *kami* of foods, Wakaukanome, and carried out rituals and prayed for the success of rice paddy planting. At the same time, Tatsuta Taisha carried out the *kaze-no-kami matsuri* ritual and appeased the wind.

■ Yamato river

Center
Hirose Ooimi Jinja, enshrined Wakaukanome, the *kami* of foods. Rivers flow from surrounding mountains and join in the center of the basin. The *jinja* is at the merging point. The joint service for all *kami* is conducted here.

Kami of Mountain
Mikumari: *kami* of mountains, enshrined *kami* of the water source.

Kami of Water
Yamaguchi: *kami* of water, enshrined *kami* of streams.

Kami of Rice Fields
Miagata: *kami* of rice fields, enshrined *kami* of agriculture estate.

Kami of Wind
Enshrined *kami* of wind. Protection from strong winds.

A *Jinja* cluster in Nara basin and the surrounding mountains. *Jinja* with enshrined *kami* of water, mountains, rice fields and wind are stationed at each corresponding essential location.

Kumano↓

YAMATO BASIN 大和盆地

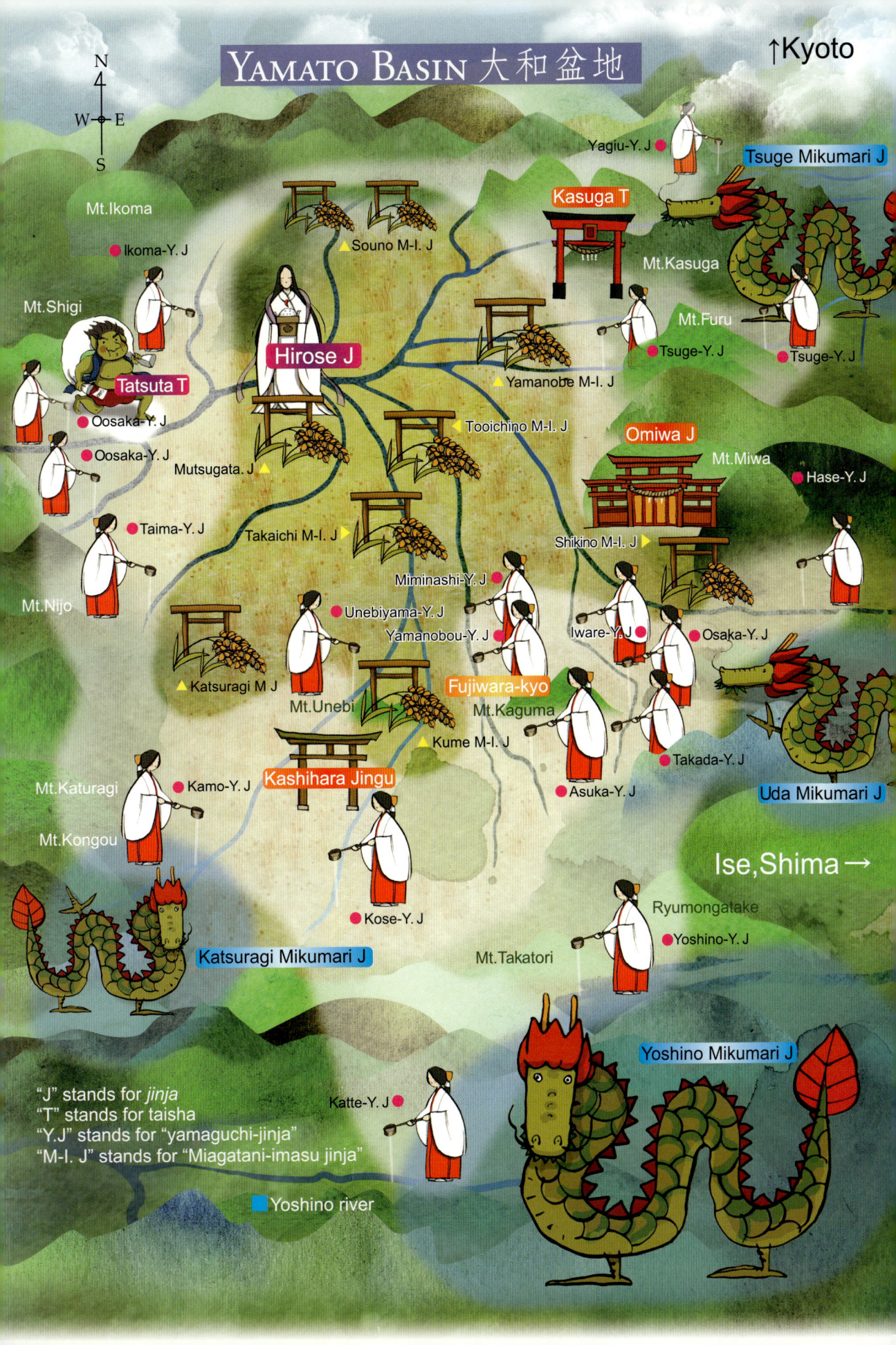

↑Kyoto

N W E S

- Mt.Ikoma
- Ikoma-Y. J
- Souno M-I. J
- Yagiu-Y. J
- Kasuga T
- Tsuge Mikumari J
- Mt.Shigi
- Mt.Kasuga
- Tatsuta T
- Hirose J
- Mt.Furu
- Yamanobe M-I. J
- Tsuge-Y. J
- Tsuge-Y. J
- Oosaka-Y. J
- Oosaka-Y. J
- Mutsugata. J
- Tooichino M-I. J
- Omiwa J
- Mt.Miwa
- Hase-Y. J
- Taima-Y. J
- Takaichi M-I. J
- Shikino M-I. J
- Mt.Nijo
- Miminashi-Y. J
- Unebiyama-Y. J
- Yamanobou-Y. J
- Iware-Y.J
- Osaka-Y. J
- Katsuragi M J
- Mt.Unebi
- Fujiwara-kyo
- Mt.Kaguma
- Kume M-I. J
- Takada-Y. J
- Kashihara Jingu
- Asuka-Y. J
- Mt.Katuragi
- Kamo-Y. J
- Uda Mikumari J
- Mt.Kongou
- Ise, Shima →
- Ryumongatake
- Yoshino-Y. J
- Kose-Y. J
- Katsuragi Mikumari J
- Mt.Takatori
- Yoshino Mikumari J
- Katte-Y. J

"J" stands for *jinja*
"T" stands for taisha
"Y.J" stands for "yamaguchi-jinja"
"M-I. J" stands for "Miagatani-imasu jinja"

Yoshino river

Yoshimuta Headwaters

吉無田水源

The Yoshimuta headwaters are located on the west side of the outer rim of Mt. Aso, Kumamoto prefecture. The outer rim of Mt. Aso is located on volcanic plateaus and one of its features is its lack of forest. Because of this lack of forest, Nanataru village (currently named Mifune-machi, Kumamoto prefecture), located at the foot of the outer rim of Mt. Aso, was short on water resources. For more than half a century during the Edo period, the villagers continuously planted trees in order to secure a water source. The project began in Bunka 12 (1815), and a total of 2,400,000 trees were planted until Koua 4 (1847). The tree-planting continued until Keiou 3 (1867).

The water resulting from the tree-planting still supports the lives of people living downstream.

Left photo: Mizu Jinja and right photo: Yama Jinja located by the river
People who come to get water offer prayers of gratitude to *kami* before returning home. The water has been chosen as one of the 100 best mineral waters in Kumamoto.

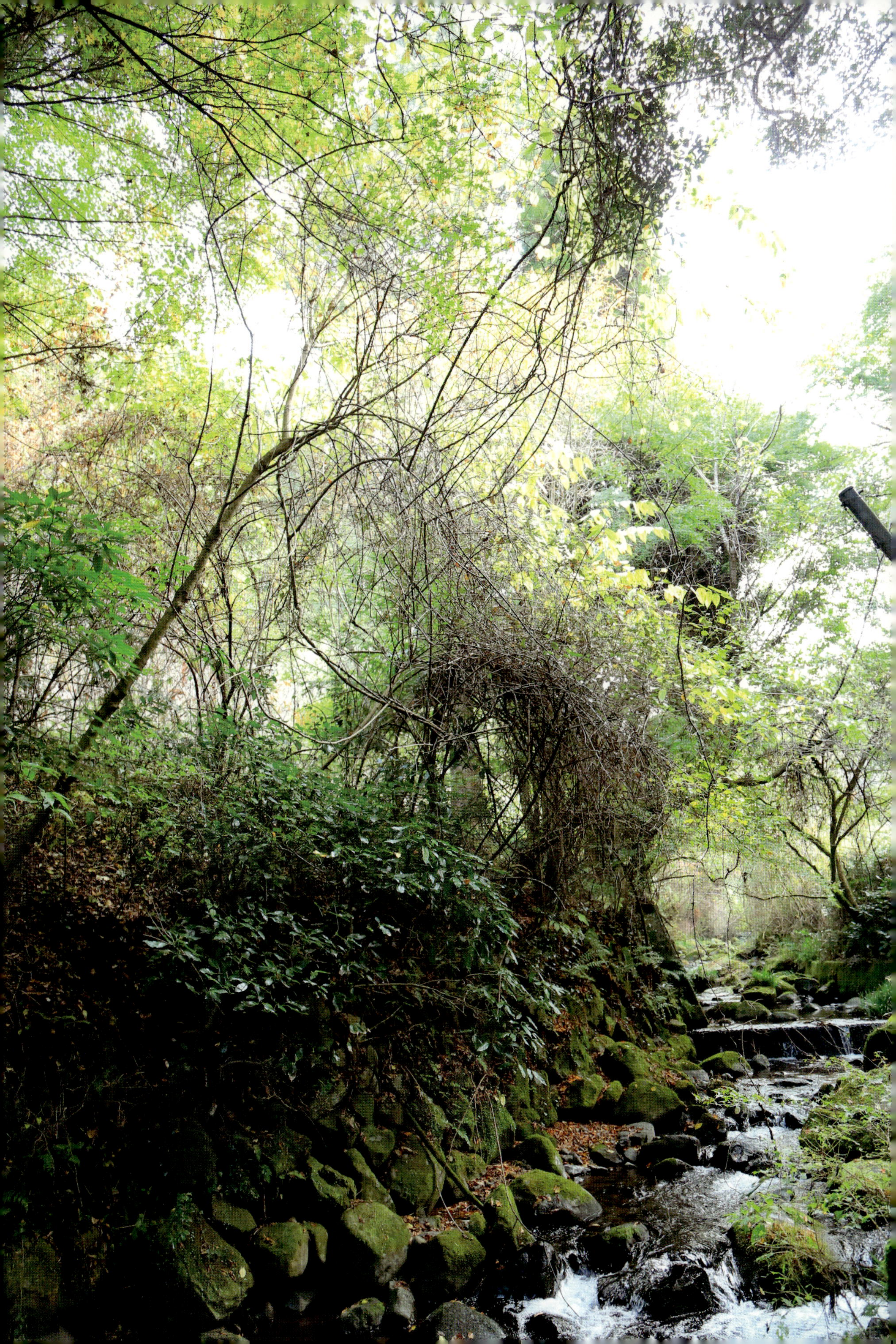

Enshrined Water Resource: Kifune Jinja

水源神を祀る：貴船神社

An interview with Kazuhiro Takai 高井和大
by Miki Miyake 宮家美樹

In the mountains of northern Kyoto, there is a *jinja* that people have revered as a water provider *kami* since the time Kyoto was the capital of Japan. In the *jinja*, rain ceremonies were held with the emperor's contribution. A black horse was offered during times of drought and a white horse was offered during periods of long rain. The *jinja*'s name is Kifune Jinja. Its origin is unknown. The main *jinja* was originally an *okumiya* (innermost *jinja*), but flood damage forced its relocation to its current location in 1055 (Tenki 3, middle of the Heian period).

Kifune Jinja is located by the Kibune river which is the flow source to the Kamo river which flows through the middle of Kyoto city from north to south. I asked the head priest of Kifune Jinja, Kazuhiro Takai, to talk about "Water, Mountain and Forest".

"Kifune Jinja's enshrined *kami* is called 'Okami'. Long ago, *Okumiya* enshrined 'Kuraokami' and Hongu(main sanctuary) enshrined 'Takaokami'. Currently they are unified as 'Okami'. The *jinja* history states that 'they are different by name but are the same *kami*.'

'Okami' means a place from which water is emerging. The 'Taka' in 'Takaokami' means 'high', so 'Takaokami' means the *kami* of water emerging from a high place. In front of the main building of Hongu, the sacred water emerges through the rocks on the side of a cliff, which is a high place. In Okumiya, it is said that there is a dragon hole (a sacred cave where a dragon *kami* lives) underneath the main building. Water most likely used to be gushing from that hole. 'Kura' in 'Kuraokami' means 'dark'. Water was gushing from the deep and dark bottom of the gorge, so it is called 'Kuraokami'.

The Chinese character of 'Okami 龗' consists of three parts: 雨 (*Ame*: rain), 口 (*Kuchi*: mouth) and 龍 (*Ryu*: dragon). The three 口 between 雨 and 龍 represents vessels to put food offerings for *kami*. *Kami* of water is sometimes called *Ryujin* (dragon *kami*). The Chinese character of 'Okami 龗' shows that people used to pray for rain by giving offerings to the dragon *kami*.

Kifune jinja 貴船神社
180 Kurama Kibune-cho, Sakyo-ku Kyoto-shi, Japan
Right photo: Stone steps heading to the main building

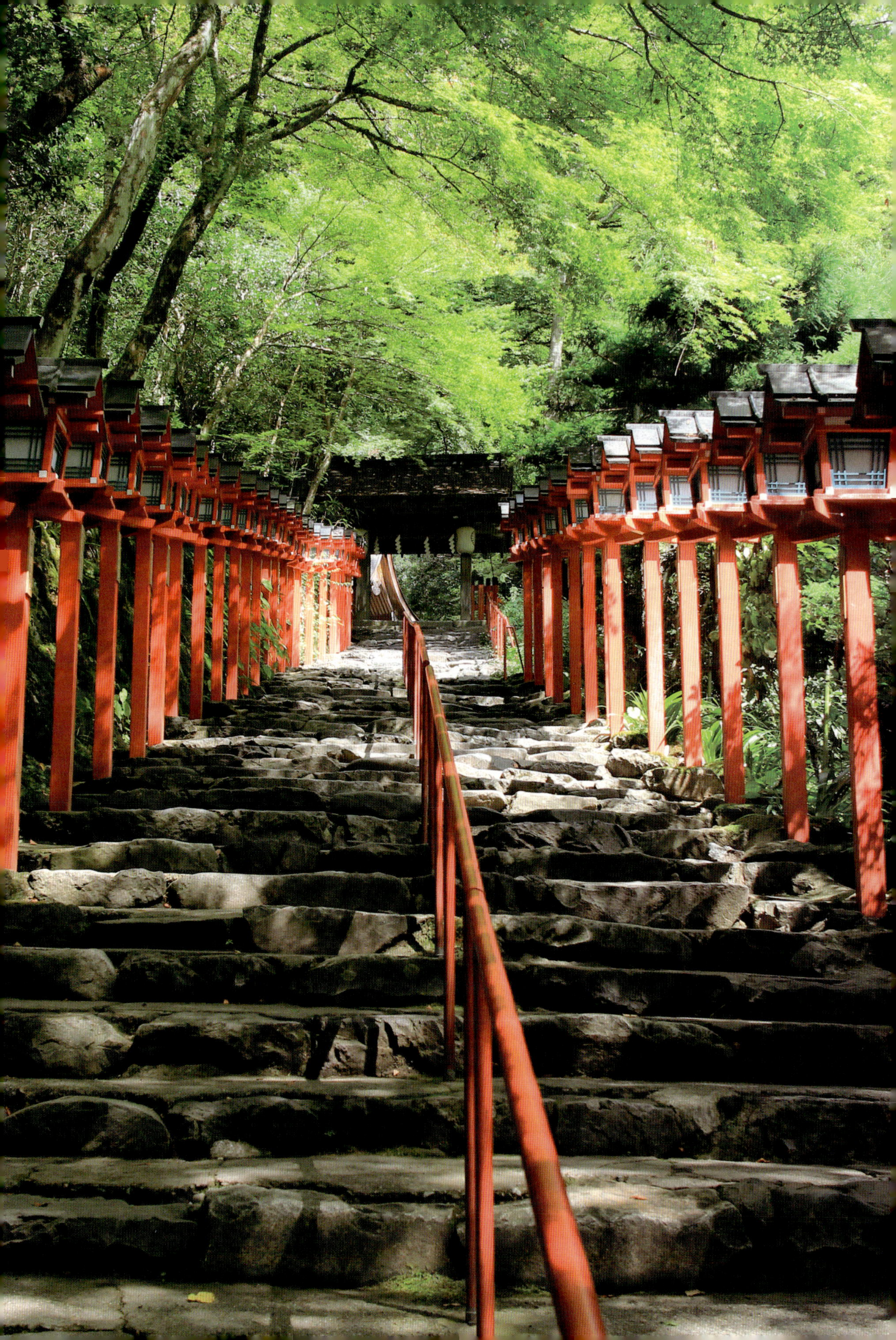

Chief Priest Kazuhiro Takai 高井和大

Participated in the 3rd World Water Forum and delivered keynote speech with architect Tadao Ando.

In Japan, there is a *kami* of river in the river and a *kami* of ocean in the ocean. Even a *kami* of the mouth of the river, called Seoritsuhime, exists. Mizuhame-no *kami* is another typical *kami* of water. Among various *kami* of water, 'Okami' would be the *kami* closest related to the lives of humans.

The word '*Kawakami*', meaning 'upriver', contains the sound of the word '*kami*'. Ancient people believed a special power existed upriver. What is that power? In ancient Japan, people believed *kami* descended from above. The *kami* descending is water. Water, the source of life, falls down from the mountains. Water descends continuously from the deep mountain and dark gorges upriver. This must be a blessing from the *kami* of water upriver. The ancient Japanese may have had this way of thinking. Our ancestors felt some great power although they did not quite understand it. Although they didn't say it in words, everybody carried an awareness of the inexplicable power which is currently recognized by the Japanese as *kami*. From here on, this is my own understanding: the great inexplicable power will be the *kami* of Kifune. In other words, the *kami* of Kifune is the origin of the word *kami*.

The mountains contained the spirit world that humans cannot enter easily. The water that sustains our lives flows from the mountains. It is not hard to imagine that fear and awe towards the mountain became the basis for the idea of *kami*."

About 1 million people die each year because of lack of water. Considering this, Japan can be said to be a most blessed nation with its water resources because the Japanese have treasured the forests for a long time.

Japan, as a nation of 'Wood Culture', used copious amounts of wood for housing materials and other uses, so a large amount of wood was logged. However, Japan's forest area still covers 70% of the land. I think this is because the Japanese knew that conserving forests also lead to the conservation of water. Herein lies the origin of the Japanese philosophy of revering nature as *kami*.

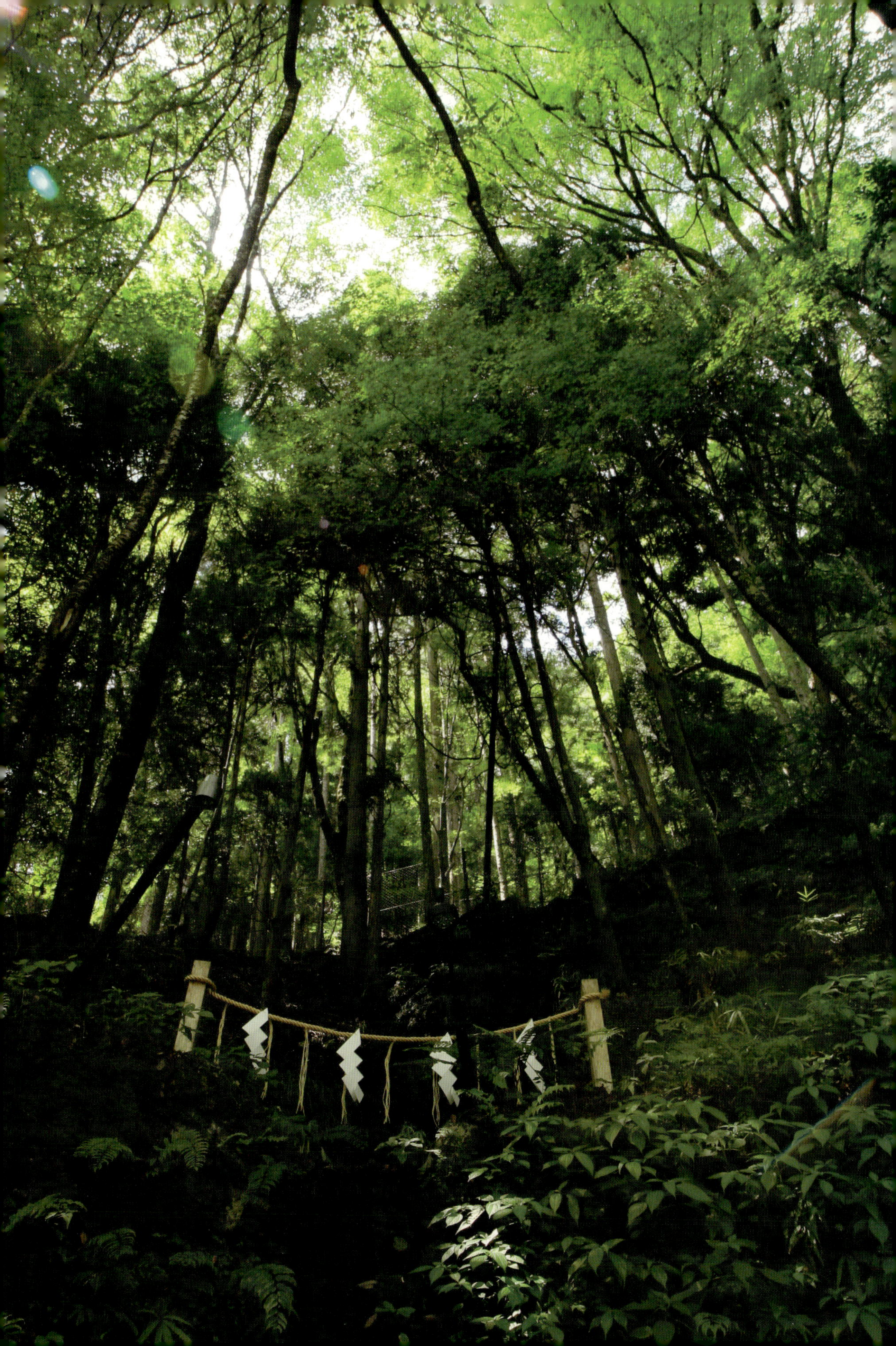

Rainfall itself is not enough to supply water. The rain has to be stored and allowed to emerge little by little. The fallen leaves in the mountain make spongy soil through which trees firmly spread roots. These roots keep water deep underground and 100 to 200 years later, the water emerges as a river and becomes a water supply to support the lives of people. Kibune river is the water source of Kamo river and Yodo river (the river that flows into Osaka bay). The stream starts from Seryo peak, 4 km upstream, and flows north, becoming Katsura river, which flows toward the Arashiyama area of Kyoto. Kibune is the original stream, so when there is heavy rain, the water level suddenly increases. It has caused serious flood damage before, but during dry weather, the water volume is the same as usual. Children often ask me where this water comes from. This 4-kilometer wide forest generates the rich water. It is really a marvelous thing.

THE exquisite balance of rain, earth, and trees produces the naturally circulated 'precious water' used by humans to relieve one's throat and to farm rice patties. The appreciative attitude towards trees, forests, and mountains is comparable to praying to the *kami* of water.

Japanese took good care of trees and treated them with care. Before cutting trees, they had rituals underneath the tree and said '*Itadakimasu*', which meant that they will humbly receive and consume a life. In addition, they planted twice the amount of trees for every tree cut down. They had a wisdom about planting trees.

I don't have a concrete plan to preserve forests and water. However, I think the most important thing is that people have the mindset that 'the environment must be preserved'. With that point of view, I think our *jinja* is closely related to this matter, since within our *jinja* is enshrined the water source that is most closely located to the residential area."

Priest Takai is currently participating in the "Koji no mori" (traditional forest) project that raises large-diameter trees. The project raises trees for 300 to 400 years to supply lumber for *jinja*, temples, or other cultural asset buildings. This will be an act of forest conservation as well. The wisdom that the priest carries on from ancestors must be passed on to future generations through such activities. To do so, we all must do our part.

Temizuya by the main building. Sacred water flows from Mt. Kifune.

Everything is Alive

Shinto priest, Translator
Tomohito Sunami 須浪伴人

It is necessary to begin by clarifying that *Shinto* is not a faith solely focusing on nature. *Shinto* has been drawing recent attention from abroad for its 'nature-centric' character. Magazine articles, television programs, and internet websites are discussing *Shinto* as if it were a 'newly discovered idea to save nature'. To some extent, this interpretation of the Japanese traditional faith is acceptable for *Shinto* priests since *Shinto* does have a concept of worshipping nature as *kami* (deity). Indeed, there are many *Shinto* shrines which enshrine the *kami*; the *kami* of rain, the *kami* of wind, the *kami* of water, and so on. However, in addition to these *kami*, Japanese people have also worshiped human beings who have made great achievements. For example, Sugawara Michizane (菅原道真) who was an ancient scholar, poet and politician is enshrined in Tenmangu and is now worshipped as a *kami* of scholarship for his intelligence. You will also find Toshogu shrines almost everywhere in Japan and these enshrine Tokugawa Ieyasu (徳川家康), a founder of the Tokugawa shogunate. There are many other similar examples in Japanese history, and such a form of enshrinement is an important factor of *Shinto*.

To understand this unique characteristic of *Shinto* enshrinement, one may need to grasp the *Shinto* narrative of how the world was conceived. In the story of the divine age, compiled in the Kojiki and Nihonshoki, creation of the main Japanese islands is narrated as a work of the divine couple of Izanaki, a male deity, and

Izanami, a female deity. Using a sacred spear, the two *kami* first create the Japanese archipelago from the ocean of chaos. Many myths from other faiths and traditions have similar stories of creation but the remarkable point of Japan's story is the way in which it has been named. This divine activity of creation is called kuniumi, "birth of the country". As its name indicates, it is not a mere creation of things but a creation of life. After the Japanese archipelago was created, many other elements of the world were born from the divine couple. Therefore, every single existence in this world, including non-living things, is believed to be alive and has life force within it. From this perspective, human beings form only a part of nature and there is no distinct hierarchy between living entities in this world.

Balance is Important

Today, many people believe that we are ruling the world and can conquer nature by the power of science and technology. In the last century, many forests disappeared because of clear-cut logging, and rivers and oceans were polluted by industries. These trees have been used as building materials and fuel to help people 'develop and prosper', and materials produced by industries have made our life more comfortable. People gradually started to think that they can not only overcome but also can conquer nature. However, human activities cannot go any further beyond the limit set by nature. This is evidenced by looking at tragedies caused by natural disasters. There are typhoons, earthquakes, droughts, floods, cold waves, to name a few, but no one has been successful warding off such natural disasters. Japanese people were certainly forced to realize that human beings are helpless before the great power of nature when they experienced the earthquake in 2011.

Today, our society is facing environmental issues. Global warming is probably the most discussed issue, and as of yet there is no clear solution. Scientists and scholars claim that activities of human beings have exceeded the maximum capacity of nature and that destruction of our natural environment has caused such climate change.

The argument is relevant to our larger point. Before the appearance of the first human beings in this world, air, water, trees and other elements of nature were already there. Animals eat vegetables, return to the earth when they die, and then trees grow on that earth. This chain was already formed before the history of human beings. Hence, balance within nature was well maintained before humankind spread to every single corner of the earth. A group of herbivores died of hunger because they had eaten all the leaves of a nearby forest, thereby killing the trees. Trees that grew too much shut out sunlight and thereby killed smaller plants. With the exception of human beings, all living things in this world have tried not to exploit another in order to avoid extinction. They knew instinctively that all living things are interdependent.

In order to keep nature in good condition, therefore, it is essential to maintain the balance between its protection and utilization. It might be too late to get rid of what we have materialistically developed and gained through history. It would be impossible to live without using natural resources. For example, night would be total darkness without electricity and if one wanted a light, he or she would have to cut a branch to burn. On the other hand, it is also true that we cannot exploit nature anymore. Scientists have proven that deforestation in mountains causes the reduction of running water in rivers, and ultimately a reduction in ocean levels and the amount of fish in the oceans. As world climate change continues to be discussed, it is also vital that we argue for protecting forest, river and other natural environments for all who live on this planet.

SHINTO IS APPRECIATION

Human beings cannot survive without sacrificing other forms of existence. We have to eat meat and vegetables to live. Although some people are vegetarians in order to avoid taking lives of other living existence, this experience is not true for everyone. From *Shinto*'s perspective as described previously, vegetables are also alive and eating them means taking their lives as part of the chain of life. Everything in this world is alive, and every single life can be worshiped as *kami*. Things are interdependent and humankind is not the supreme form of existence.

What is important in *Shinto* is that whenever people take something from nature, it has to be done with respect and appreciation. As long as this is not forgotten, people will be able to keep a good balance between development and conservation.

SACRED MOUNTAINS
聖なる山

During ancient times in Japan, there were no sacred buildings for *jinja*. Instead, Japanese people revered forests, rivers, rocks, and the ocean in the pure natural world. Among the revered nature were sacred mountains. There are usually *jinja* at the foot or summit of a sacred or well-known mountain. These *jinja* enshrine the mountains. For thousands of years, Japanese people have offered prayers of gratitude toward the *kami* of water for supplying rich water at these *jinja*, and they treasured and took good care of the forests.

Here, we introduce these *jinja* which enshrine the mountains.

photo:pixta

Jinja building at sumit of Mt. Fuji

Mai *jinja* building at Fujinomiya city

Mt. Fuji: Fujisan Hongu Sengen Taisha
富士山・富士山本宮浅間大社

Main *kami* enshrined: Asama-no-ohkami
(also known as Konohana-no-sakuyahime-no-mikoto)

Mythology: Konohana-no-sakuyahime-no-mikoto is said to be a very beautiful *kami*. She married Ninigi-no-mikoto who descended from Takamagahara (dwelling place of *kami*). Soon after the marriage, she became pregnant, but the husband questioned her pregnancy. In order to prove that the baby was her husband's child, Konohana-no -sakuyahime-no-mikoto went into the delivery room without a door. She set fire to the room and vowed that if the child were truly the offspring of Ninigi-no-mikoto, then it would not be hurt. She gave birth in the flames and safely had three babies.

Mt. Fuji is famous as the symbol of Japan. It is the tallest mountain in Japan and its breathtaking figure receives praise as the world's most beautiful mountain. *Okumiya* (innermost *jinja*) is at the top of Mt. Fuji and *Hongu* (main *jinja*) is at the southwest foot of Mt. Fuji. Mt. Fuji was registered as a world heritage in Heisei 25 (2013).

According to the *jinja*'s records, Mt. Fuji had a large eruption during the reign of the 7th Emperor Korei. Most residents from around Mt. Fuji moved away and the Mt. Fuji area was isolated for a long period. This worried the 11th Emperor Suijin, and led him to enshrine Asama-no-ohkami (Konohana-no-sakuyahime-no-mikoto) in 27 BC to appease the mountain *kami*. By the power of Konohana-no-sakuyahime-no-mikoto, volcanic activity was reduced and peaceful days returned.

Currently, Fujisan Hongu Sengen Taisha is the head *jinja* among 1,300 Sengen *Jinja* throughout Japan. In the courtyard of Hongu, there are 500 sacred cherry trees and it is a popular place in the Spring to view cherry blossoms. Mt. Fuji is open to hikers in July and August only, and during this time, Shinto priests stay in Okumiya and offer prayers. About 300,000 people climb Mt. Fuji per year.

Active Volcano located in Shizuoka and Yamanashi prefecture
Elevation: 3,776meters singular peak
Address: 1-1 Miyamachi, Fujinomiya city, Shizuoka prefecture, Japan

Sando toward main jinja building

Main jinja building

Mt. Hakusan: Shirayama Hime Jinja
白山・白山比咩神社

Main *kami* enshrined: Shirayamahime-no-ohkami

Mythology: Shirayamahime-no-ohkami, also known as Kukurihime-no-kami, is well known as the *kami* who acted as a mediator between Izanagi-no-mikoto and Izanami-no-mikoto when they had an argument, as mentioned in the Nihon Shoki (The Chronicles of Japan, the book of classical Japanese history). Izanagi-no-mikoto and Izanami-no-mikoto are *kami* who created the land of Japan and other countless *kami*. In the name Kukurihime-no-mikoto, "*Kukuri*" stands for "*kukuru*", which means "to tie up". She is revered among Japanese as the *kami* with the power to bind or join together and create harmony.

Okumiya of Shirayama Hime Jinja is located at the summit of Mt. Hakusan and *Hongu* is located alongside the Tedori River, fifty kilometers below Mt. Hakusan. Entry to the mountain was once prohibited as it was recognized as a dwelling place of *kami*. In Yoro 1 (717), Monk Taicho was the first to climb Mt. Hakusan to pray.

Shirayama Hime Jinja's establishment dates back to Suijin 7 (91 BC). In the long history of the *jinja*, *Hongu* changed its location three times until settling at its current location in Chokyo 2 (1488). People were able to survive by receiving water from the Tedori River, through which the melted snow of Mt. Hakusan flows. In appreciation, people offered prayers of gratitude to Mt. Hakusan, which is the origin of prayer to Mt. Hakusan. Mt. Hakusan is a national park and also known for alpine flora treasury. At Murodo-daira, located south of Gozengamine (the tallest peak of Mt. Hakusan), there is a *Kito-den* (prayer hall) and *Sanro-den* (private prayer hall). During the summer (July and August), when the mountain is open, priests and maidens stay at Murodo-daira. At sunrise, they yell "*banzai*" three times for world peace, and then, conduct a daily service in front of the Okumiya building.

Mt. Hakusan spans over four prefectures: Ishikawa, Fukui, Gifu and Toyama.
Elevation: Gozenga Peak 2,702 meters, Ohnanji Peak 2,684 meters, Mt. Bessan 2,399 meters, Kengamine 2677 meters, Sannomine 2128 meters. These five mountains are called "Hakusan five peaks" and Hakusan is the general term for the surrounding mountains.
Address:105-1 Ni, Sannomiya-machi, Hakusan city, Ishikawa prefecture

The ritual is performed at sumit on every May 10

Main jinja building

The Yahiko Mountains: Yahiko Jinja
弥彦山・彌彦神社

Main *kami* enshrined: Ameno-kagoyama-no-mikoto

Mythology: Ameno-kagoyama-no-mikoto is referred to as Kumano Takakuraji in the Nihon Shoki, and a historical Japanese text Sendai Kujihongi refers to him as a great grandchild of Amaterasu. Ameno-kagoyama-no-mikoto received an imperial decree four years into enthronement of the first emperor Jinmmu, to open up Echigo (currently "Niigata"). He landed on Nozumihama, at the foot of Mt. Yahiko on the Japan sea side, and introduced rice farming, silkworm culture, fishery, salt manufacturing, and sake brewing, and put efforts into industry development in Echigo.

Mt. Yahiko is called "the first mountain to be hit by the morning sunlight" in the Echigo area. The main *jinja* building is located at the eastern foot of Mt. Yahiko. The *goshinbyo* (a burial place) of Ameno-kagoyama-no-mikoto and his wife is located at the summit.

From the summit of Mt. Yahiko, one can see the Echigo plains to the east and the Japan sea to the west. It is as if the *kami* couple have been watching over the industries of Echigo together for thousands of years. Before the energy revolution, Mt. Yahiko played a role as "*Satoyama*", a backyard woodland, and provided essential firewood. Mt. Yahiko can be seen from both the ocean and the land, so it has also served as a landmark since ancient times. It is said that the ten treasures that Ameno-kagoyama-no-mikoto brought back from Yamato (currently "Nara") are buried in Mt. Tahou, next to Mt. Yahiko.

Niigata Prefecture
Elevation: The Yahiko Mountains
North to South: Mt. Kakuda-482meters, Mt. Taho-634meters, Mt. Yahiko-634meters, Mt. Kugami-313meters
Address:2887-2 Yahiko, Yahiko-mura, Nishikanbara-gun, Niigata Prefecture

Main jinja building taking sunset light.

Beautiful Sando

Mt. Houman: Houmangu Kamado Jinja
宝満山・宝満宮竈門神社

Main *kami* enshrined: Tamayori-hime-no-mikoto

Mythology: Tamayori-hime-no-mikoto is a sister of Toyotama-hime. After Toyotama-hime's death, her son, Ugaya-fukiaezu-no-mikoto, was left to be raised by Tamayori-hime-no-mikoto. Tamayori-hime-no-mikoto then also gave birth to Wakamikenu-no-mikoto, who later became the first emperor Jinmu.

Mt. Houman is located at the center of Fukuoka plain of Kyushu and in the *kimon* direction (northeastern direction, devil's gate) of Dazaifu, the western capital of Japan that at one time controlled the entire Kyushu area. From the summit of Mt. Houman, one can overlook the plains of Fukuoka and Dazaifu.

The origin of the enshrinement of *kami* at Mt. Houman dates back to 1,350 years ago in order to protect Dazaifu. Later, when priest Shinren shut himself in Mt. Houman for Buddhist ascetic practice, Tamayori-hime-no-mikoto appeared to him. The imperial court founded the upper *jinja* on the summit of Mt. Houman to venerate this site, and this is the foundation of Kamado Jinja.

There is a Jogù (upper *jinja*) at the summit and Gegù (lower *jinja*) at the foot of Mt. Houman. Dazaifu played an important role in diplomatic relations and functioned as a window of international relations to take care of welcoming foreign guests or sending Japanese ambassadors abroad. Japanese envoys to China climbed Mt. Houman before their voyage and looked in the direction in which they were going, praying for safety and protection from enemies they might encounter during their voyage. After a long and important history, Mt. Houman has been revered as the mountain where *kami* resides.

On October 17, 2013, Mt. Houman was certified a national historic site. Kamado *Jinja* is widely known as a *jinja* enshrining the *kami* of good relations, marriage, and protection from misfortune related to direction.

Fukuoka prefecture
Elevation: 830 meters
Address: 883 Uchiyama, Dazaifu city, Fukuoka prefecture

Sacred Mountain Location Diagram

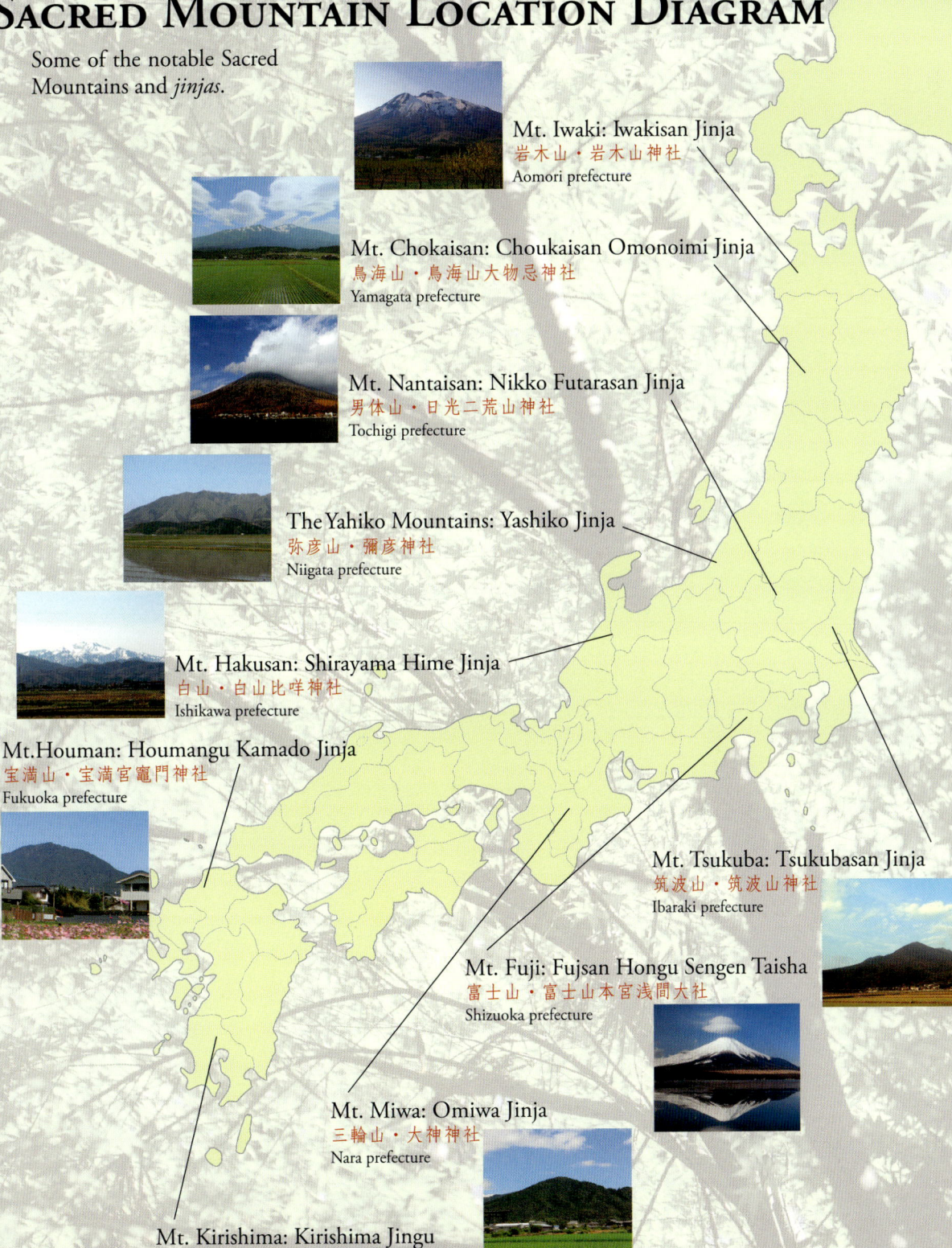

Some of the notable Sacred Mountains and *jinjas*.

Mt. Iwaki: Iwakisan Jinja
岩木山・岩木山神社
Aomori prefecture

Mt. Chokaisan: Choukaisan Omonoimi Jinja
鳥海山・鳥海山大物忌神社
Yamagata prefecture

Mt. Nantaisan: Nikko Futarasan Jinja
男体山・日光二荒山神社
Tochigi prefecture

The Yahiko Mountains: Yashiko Jinja
弥彦山・彌彦神社
Niigata prefecture

Mt. Hakusan: Shirayama Hime Jinja
白山・白山比咩神社
Ishikawa prefecture

Mt. Houman: Houmangu Kamado Jinja
宝満山・宝満宮竈門神社
Fukuoka prefecture

Mt. Tsukuba: Tsukubasan Jinja
筑波山・筑波山神社
Ibaraki prefecture

Mt. Fuji: Fujsan Hongu Sengen Taisha
富士山・富士山本宮浅間大社
Shizuoka prefecture

Mt. Miwa: Omiwa Jinja
三輪山・大神神社
Nara prefecture

Mt. Kirishima: Kirishima Jingu
霧島山・霧島神宮
Kagoshima prefecture

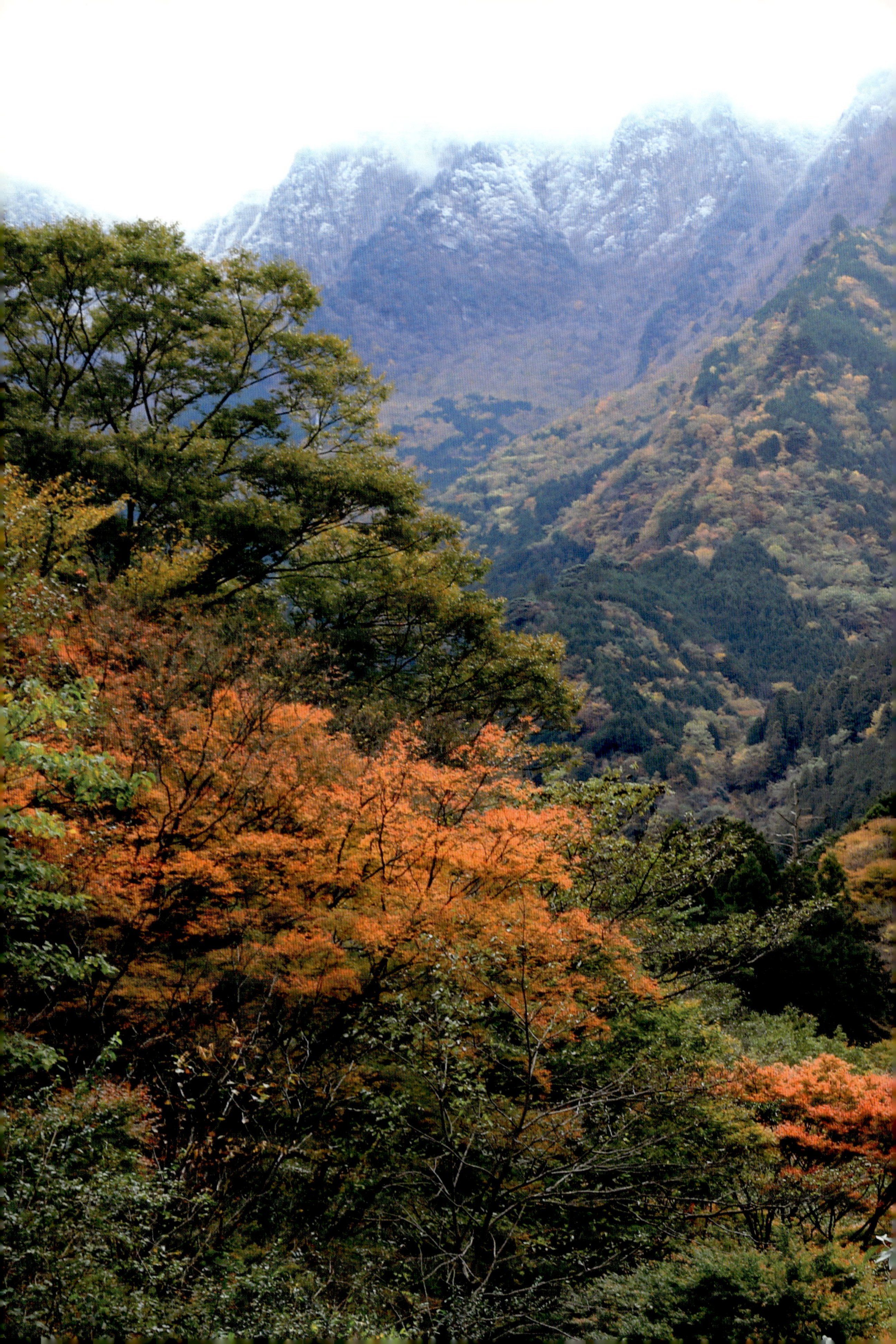

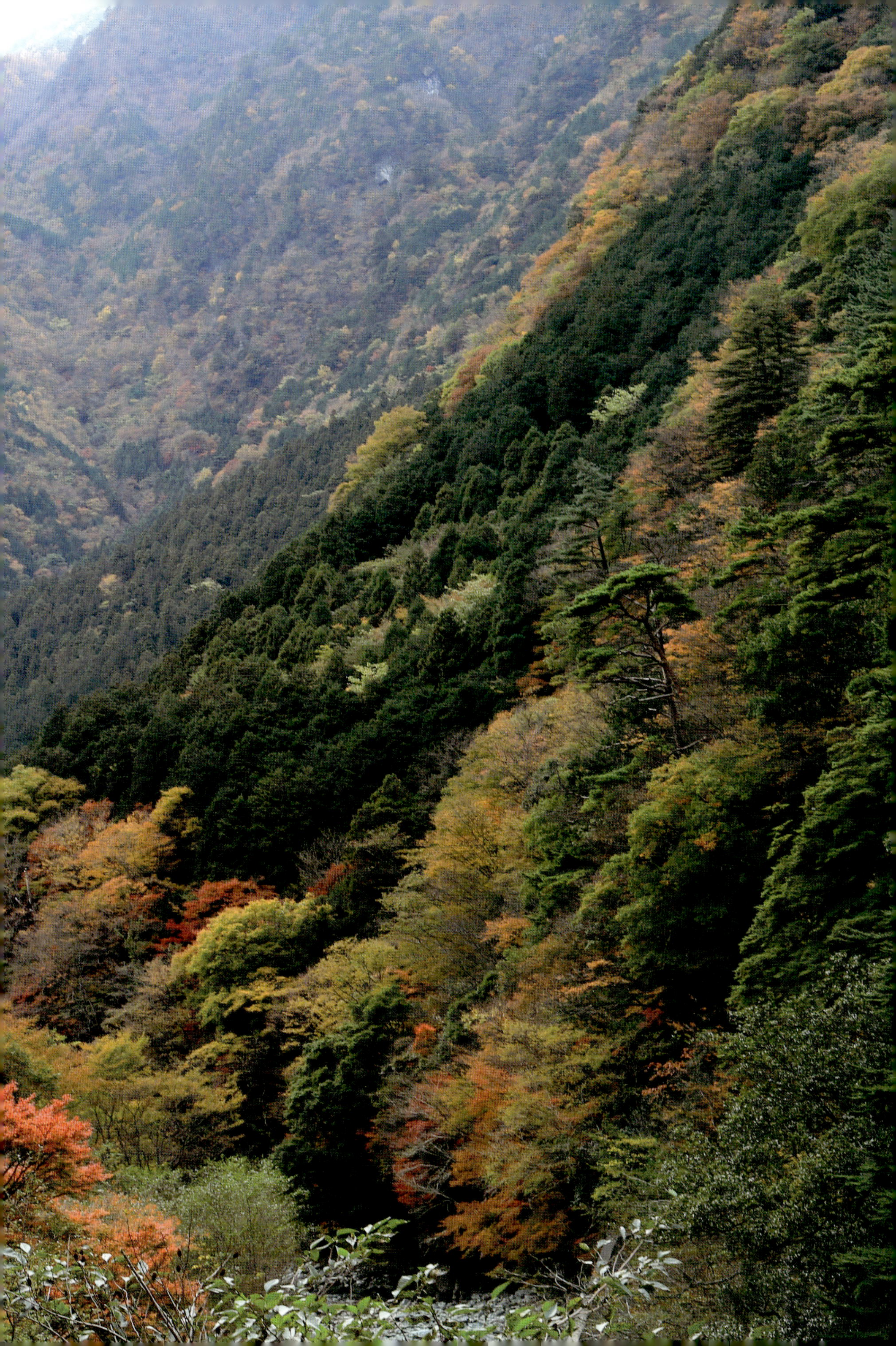

Yaegaki Jinja: swept away by the super-tsunami triggered by the Great East Japan Earthquake of March 11, 2011
八重垣神社

A Step at a Time Towards Reconstruction
再建に向けて一歩づつ！

Yaegaki Jinja is located in Yamamoto-cho, Watari-gun, Miyagi-prefecture, five hundred meters inland from the ocean. The big tsunami caused by the Great East Japan Earthquake on March 11, 2011 inflicted devastating damage upon all but two of the three hundred *ujikos*' (parishioners of a *jinja*) houses in the Kasano and Shinhama area. The main building of Yaegaki Jinja disappeared completely and its *torii* gate was demolished, leaving behind nothing but a *Mikoshi*, a portable *jinja* where *kami* are carried in a procession during festivals. The *Mikoshi* would have also been swept away, but it got stuck to the house of the head of the *ujikos*.

Two years have passed since the earthquake. At Yaegaki Jinja, new trees have been planted and now they have grown to about one meter tall, surrounding *hokora* (a miniature *jinja* structure). A refreshing breeze from the ocean blows through the *jinja*. With no surrounding buildings or big trees, the sky seems vast.

To learn more about the tsunami and Yaegaki Jinja, we interviewed Shoko Fujinami, the chief priest of Yaegaki Jinja.

Trees here are one year old since they were planted.

Right photo: Shoko Fujinami, the chief priest of Yaegaki Jinja

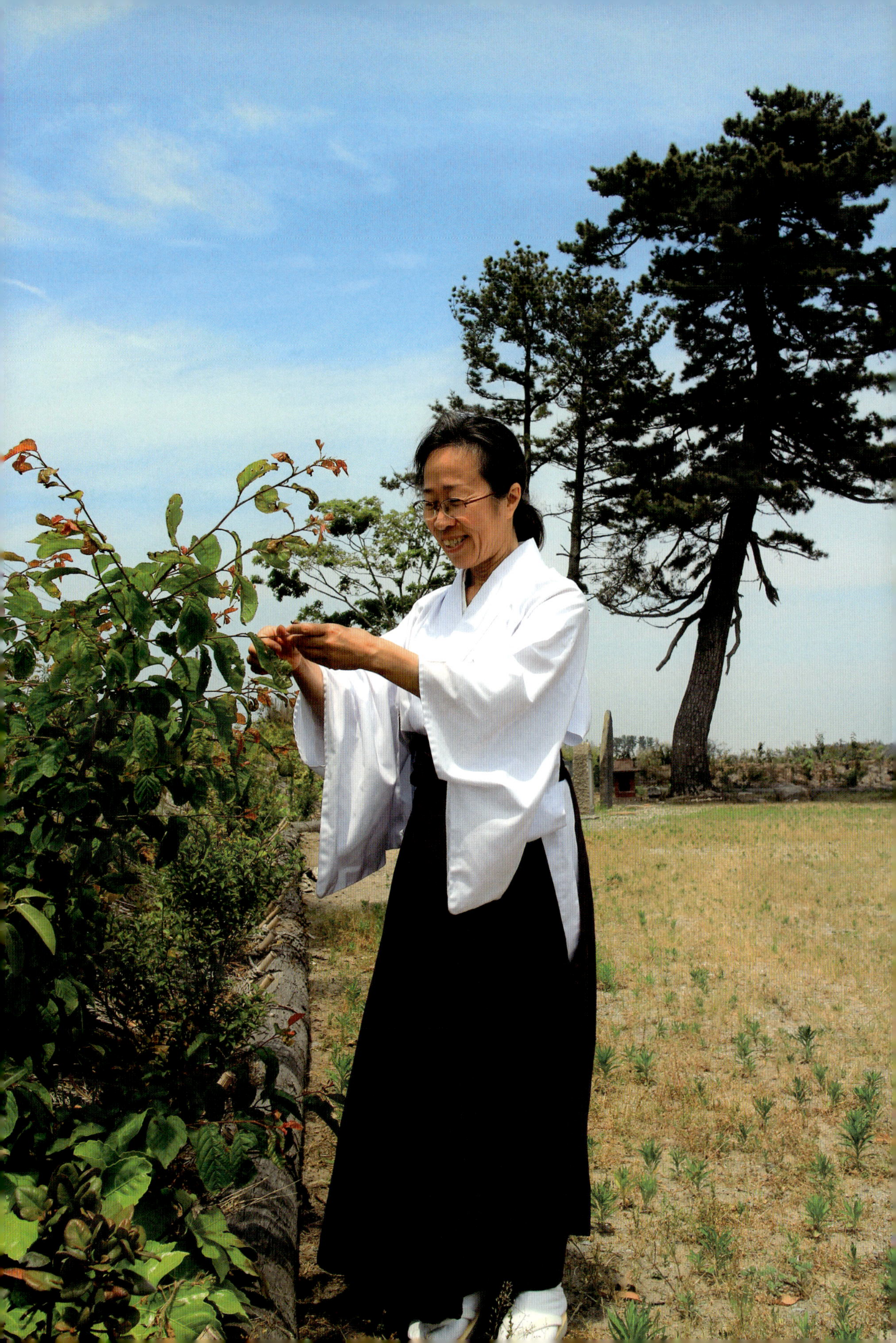

Everything was washed away at Yaegaki jinja. Small *jinja* (hokora) was sitting where the main building used to be.

Interviewer: "This area was completely washed away and nothing was left behind after the tsunami, is that right?"

Priest Shoko Fujinami (referred to as Fujinami from here on): "That's right, but many farmers who lived around here have strong beliefs carried on from their ancestors, and they neither curse nor blame nature and the *kami*. On the other hand, there are those who live on highlands and typical white collar workers who see this terrible devastation and say, "What a hellish looking sight. Do the *kami* have no mercy?!" Whereas, those who suffered a great deal of damage from the tsunami feel just the opposite; they want to be closer to the *kami*."

Interviewer: "So they have a completely opposite perspective then."

Fujinami: "When I visited the evacuation shelter, I told them about the others people's remarks that there is no *kami*. One of the old men replied 'For the person who says there is no *kami*, there was never a *kami* - even before disaster struck'. Then people around him started clapping in approval. [Fujinami laughed at the memory]. On the contrary, they say that they want *Ofuda** or *Omamori***. I realized that this earthquake shed light on people's beliefs. Beliefs were strengthened within those who originally had beliefs while those who never had belief remained as such."

Ofuda*: Talisman acts as a symbol of a *kami* to protect a household.
Omamori**: Protective amulet or charm.

Interviewer: "Their job requires a lot of patience, doesn't it?"

Fujinami: "Both farmers and fishermen know the benefits of nature, and at the same time they know the damage that nature can inflict upon them. This tsunami's damage was devastating, but the farmers and fishermen don't hold a grudge against nature. Of course, farmers depend on modern technology as well, but I feel that their way of thinking is different from those who don't work with nature.

As people who work in nature pray to the sun, many people hold their hands in prayer facing the *Ujigami**. They are not praying to the *jinja* building itself, but it just so happens that there is a *jinja* building there. Of course, if a *jinja* building is majestic, they would take pride in it. However, they would still bring their hands together in prayer even if there was no *jinja* building to begin with. This is why they come here, where there is nothing but ruins, and still pray to the *kami*."

Ujigami*: Tutelary deity of the villages and communities

June 24, 2012-530 volunteers participated and planted 3,300 trees. Photo by Japan Culture Promotion Foundation

Interviewer: "Wow, that is amazing!"

Fujinami: "Yes, it is. They picked up a roof tile from the ruins and placed it where the *jinja* building used to be. People leave *Osaisen* (offerings) there. Eventually I built a small prefabricated *jinja* office. When I was there, I spoke with visitors to the *jinja*. They would share their stories of the tsunami and sometimes I would end up speaking with them all day."

Interviewer: "What kind of stories did they share with you?"

Fujinami: "Everybody had amazing stories. There was one man who saw the tsunami form a whirlpool, and a car being sucked into the whirlpool. Watching this happen, he thought there was no hope for the car, but somehow the car reached the land. A man got out of the car and said 'It was very scary!'

Another person said he was carried away by the tsunami and was thrown around by the wave, but saved himself by grabbing onto a tree trunk that happened to be floating by.

Another person who was also washed away by the tsunami happened to be a surfer and was accustomed to the waves. However, even he was swallowed by the tsunami waves and kept getting submerged and resurfaced only to get submerged again. Once he got on land, his body temperature was very low and he was unconscious. Bystanders warmed him up and saved his life.

Another person said he was holding the hands of family members but couldn't hold on because of the strong current.

There was also a child who climbed and held onto a small tree only about 10 cm in diameter. She looked down to see other people holding on to the same tree. She was very frightened and worried whether the tree was sturdy enough to hold so many people at the same time. It was such a terrifying experience for her that she does not want to come back to this place."

Interviewer: "Hearing these stories makes me absolutely speechless."

Fujinami: "As people were sharing their stories, one of the visitors offered flowers. When the *jinja* building was there, nobody offered flowers, so I asked the visitor why she brought flowers. She replied 'Because *kami* lost his house just like us and I feel sorry for him, so I offered flowers.' It is then that I realized that the Japanese people hold their *kami* very close to them and it lives within their heart. It is not an absolute *kami*. I think it is very Japanese-like and a nice thing to have the feeling of sympathy towards *kami*.

Interviewer: "When was Yaegaki Jinja established?"

Fujinami: "It was Daido 2 (807 AD). There was a time in the *jinja*'s history that nobody took care of the building; it had even been washed away by a tsunami before. My ancestors originally came from Kyoto to inspect Tohoku area about five hundred years ago and then decided to stay here to live. The main building seemed to be built 20 to 30 years after Ansei 2 (1855AD) but we don't know exactly when."

Interviewer: "When the Jogan Earthquake occurred in Jogan 11 (869AD), it is said that there was a tsunami like this past one. Is there any record of it?"

Fujinami: "Actually, there is no record of it. Any records may have been carried away by the tsunami. There was another big tsunami from the Keicho Sanriku earthquake about 400 years ago in Keicho 16 (1611 AD). When I was four years old, the Chilean earthquake of 1960 occurred and I vaguely remember my house was flooded up to the floorboards. There was another tsunami after I graduated from college. I experienced two tsunamis in thirty years. As such, tsunamis are not unusual to me. People in this area had never experienced a big earthquake like this, but due to their nature, they continue to live here and grow rice despite the effects of the big tsunami."

Interviewer: "They continue to live here despite the tsunamis, and having their planted shrubs washed away."

Fujinami: "I mentioned earlier that *ujikos* have strong beliefs and I contemplated what to do about the *jinja*. I didn't know where such a big *jinja* building had gone. It just disappeared. The *jinja* structure was constructed in the old style and it was just placed upon the paving stones, so I assume the building just floated on water. I sometimes imagine it may be floating somewhere in the Pacific Ocean. My house was behind the *jinja* and it was completely washed away as well.

I learned from the news that some *jinja* contributed to society in many ways during this earthquake disasters. These *jinja* functioned as shelters and even as a mark of the border between the flooding and dry areas during the tsunami. I was thinking that we should rebuild Yaegaki Jinja on the highland so our *jinja* can help society as well for the future. But I realized we don't have the necessary land in the highland.

Concerned, the *ujikos* asked me what I will do about the *jinja*. And they say to me, 'Where else is there to go?' or 'If you move, it wouldn't even be Yaegaki Jinja anymore.'

This has been a place of sanctity for 1200 years, and I want to keep Yaegaki Jinja here for more years to come. Since the *ujikos* asked me to rebuild the *jinja* in the same place, I decided to rebuild it here again. Then we decided that our first line of action was to plant trees."

Interviewer: "Why did you decide to plant trees first?"

Fujinami: "The *jinja* building can be built within one year. Of course more elaborate ones take years to build. However, it takes 10 to 20 years to build a forest. Therefore, we started off with the forest.

First we created a forest for *kami* to dwell in and then we built a *jinja* building. Even if the building is small, as long as there is a proper forest, I believe that *kami* will be there as well. This is why forests are so important.

At that time, Jinja Honcho (Association of Shinto Shrines) announced a tree planting project sponsored by Nippon Foundation. We jumped on the opportunity."

Interviewer: "You had a tree planting last year in Heisei 24 (2012), right?"

Fujinami: "That's right, on June 24. Five hundred and thirty volunteers participated and planted 3,300 trees. There were people who came all the way from Aichi prefecture, about 580 kilometers away; there were student volunteers, those who lived in the shelters, and many people who moved to a variety of places after the tsunami, who had returned to help. I talked to several people and asked how they were doing. There was not enough time in the day to see and speak to all of them.

During the tree planting, I let everyone write a short message on a paperboard. There were messages such as "I'll come back in 10 years!" and "Good luck - I will also try hard to live another 20 years". Everyone seemed to be excited. Everybody looked so happy planting the trees.

It was a very happy occasion. Although I myself planted only one tree, it was definitely fun to participate.

When those participants return to our *jinja*, they first go to the tree they planted and make sure that it is alright. It must make them happy as if the tree is their child and they feel that they are leaving a part of themselves in our *jinja*."

Interviewer: "These *ujiko* people are very cheerful, aren't they?"

Fujinami: "I thought I shouldn't hold *matsuri** in the first year after the earthquake, so I told people that I would do just the prayer for the *matsuri*. However, about thirty people on the board of directors came for the prayer and asked me to also do the *matsuri* as well. They said that it wasn't necessary to hold the *matsuri* exactly like before, but they would like it to be as close as possible. 'Because *jinja* is a place of gathering and having fun,' they told me. 'For *obon*** and *ohigan**** we go to the graves to visit our ancestors and family members who perished during the disaster. But we need a place where we can laugh and rejuvenate our souls, so please hold the *matsuri*.'

We didn't have much money, so we decided to do whatever we could. We decided to have less fireworks on *Yoimatsuri* (small *matsuri* on the eve of the *matsuri*) and fix the damaged Mikoshi and have Mikoshi-togyo (Mikoshi parade)."

*Matsuri: sacred rituals with religious service and entertainment for *kami*
**Obon: an event to commemorate and honor the spirits of ancestors
***Ohigan: a holiday observed during the Spring and Fall Equinoxes during which ancestors are remembered

Interviewer: "Wow, that's a very heartwarming story."

Fujinami: "Although *Yoimatsuri* is a small *matsuri* just three hours long, many *ujikos* still gathered for it. It seemed like more *ujikos* came than before the earthquake. Middle school girls came dressed up in *yukata* (a casual summer kimono) with their hair up. Excited grandfathers and grandmothers came for the *matsuri*. Everybody looked so happy and seemed to have a really good time. I was truly glad that we held the *matsuri*."

Interviewer: "That's good!"

Fujinami: "There are many times in life when I was unsure of myself and *ujikos* had taught me various lessons. It is almost as if *kami* is using the *ujikos*' mouths to show me the way. It could be the *ujiko's* personality or their strong beliefs and close ties to the *jinja*.

I believe that how you were before something happens is more important than how you are after. Among people who have suffered the same disaster, those who have subconsciously lived their life with *kami*, as opposed to those who haven't, have something inside that they can count on.

Accordingly, people who have always lived their lives alongside *kami*, will still have something to count on, even after a disaster. Using that as a foundation, they can continue to live life positively.

I also believe, you can live life without faith, but when it comes down to disastrous times, there is quite a difference between those with faith and those without faith.

FORESTS HEAL PEOPLE

森は人を医す

Hideki Hirano　平野秀樹

Forest Bathing

In the fall of 2013, the world symposium of forest healing was held in Boston, in the United States, and sponsored by the Center for Health and the Global Environment, Harvard School of Public Health. A total of seven nations participated: Finland, England, Canada, Spain, South Korea, the U.S. and Japan. From Japan, Dr. Yoshifumi Miyazaki and others were invited and began the symposium with a speech regarding the effect of the forest on humans, based on medical data.

Forest therapy is an idea that originated in Japan and is "the relaxing effect in the forest as evidenced by medical data".

Although this idea has been around for just over a decade, Japan has a long history of revering forests such as "Chinju no mori" or "forest of *kami*", and those beliefs are still deeply rooted in Japanese people's minds.

Japan is the leader in research in this area and has conducted joint research with Korea and Finland. In addition, the International Society of Nature and Forest Medicine (INFOM: Chairman - Dr. Michiko Imai) was established in 2011, and their combined research has expanded to Nordic countries, England, the U.S. and Russia, totalling twelve countries.

The latest research underway is on "the difference in the physiological effects of forest therapy among individuals".[1] By walking in the forest, those with high blood pressure showed a decrease in blood pressure, and those with low blood pressure showed an increase in blood pressure.[2] There was also a difference in blood pressure and pulse rate before and after sitting in the forest. The density of immunoglobulin A (IgA) in saliva changed as well, which is also an indicator of good health. All of this data was gathered from experiments performed on 636 college students in over 50 locations throughout Japan.

Recent technological advancements in bio-analytical instruments allowed for the successful measuring of physiological changes in the forest. The relaxing effect of walking and bathing in the forest is scientifically proven to be superior to doing these activities in an urban setting.

The Forest Has Five Senses
五感をもつ

From Hokkaido to Okinawa, 57 forests across Japan have been certified after these medical experiments as "forest therapy bases". Certified forests influence our five senses and nourish our body and mind.

Forest therapy may show that forests and humans are equal. Forests have five senses as well. For example, the sense of hearing. A two-meter high plant in Xishuangbanna of China's Yunnan province visibly moves its leaves along with the songs sung by girls of the Dai people (ethnic group living in Xishuangbanna). The leaves circle slowly as if dancing with the song. This dancing grass (telegraph plant) is mentioned in the "Youyang zazu" (a miscellany of various stories and topics from the south slope of Mt. You, 860AD). Farmers in India describe this dance as a "dance that makes leaves move like snakes".

Plants also have a sense of touch. Vines or roots choose which direction to move towards with their sense of touch during the growth process. Sensitive plants or Venus flytraps move even quicker. When they are touched, they react as if they have a will of their own.

Plants and humans are creatures living in the same solar system and thus, forests also have a sense of sight. An experiment conducted by Dr. Novoplansky in 1990 confirmed that plants are able to distinguish colors. In the experiment, Verdolaga (portulaca oleracea), the weeds seen on roadsides or fields, grew while avoiding a green plastic block. Verdolaga would bump into white or red blocks, and their growth would be stunted.

Plants have a sensor to sense colors (wavelength) and have been shown to recognize green colored objects and avoid them before touching while growing. This is why leaves don't hit each other.

It can be said that plants have the sense of smell. Oak trees contain tannin and when a hairy caterpillar eats its leaves, the tree increases the tannin production. When the amount of tannin increases, the leaves become bitter so insects or animals do not eat them. When trees feel stress from their leaves being eaten or branches being broken, they increase the amount of tannin. This sociological change can somehow spread to an adjacent tree. The adjacent tree which has not yet been damaged by the caterpillar will increase tannin production. In an experiment on poplar trees in 1980, the tannin production of damaged poplar trees increased twice in 50 hours and the adjacent undamaged poplar trees increased the concentration of tannin by 60%. Some pheromone-like substances (i.e. leaf alcohol) may be a contributor to this phenomenon.

Last but not least, plants carry the sense of taste. The carnivorous sundew plants do not react to a small branch or a glass rod, but quickly react to and eat things containing nitrogen such as a fly, a mosquito, a piece of meat or hair. The Nepenthes Rajah pitcher plant can even digest a rat.

Of course, plants don't have a nerve system or brain, but they do have a corresponding system that are similar to animal muscles or nerves that react to light, electromagnetic waves, wind, and temperature change. Plants also have "impulses" which can feel stress.

I believe plants have a sophisticated communication ability and information processing system well beyond our imagination.

The Forest Heals Humans and The Nation
森は人を医し、国をも医す

The sense of time and space in the forest world is different from that of the human world. Otherwise, there is not much difference between human life and the life of forests. Some old trees have lived over three thousand years. You might think of it as forests just live slower than humans, so we don't perceive them as being "alive".

Sometimes the forest helps to solve our human problems, or to encourage us. It may be that their natural sense of time and space is what gives the forest the power of healing. If we think in this manner, the forest may make us aware of unhealthy parts of human society. "The superior doctor doctors the country, the mediocre doctor doctors people, the inferior doctor doctors disease[*3]". This is a saying from the Six Dynasties period of ancient China (222 to 589 AD), and is said to originate from "The superior doctor doctors the country, then doctors people" from "Guoyu" Discourses of Jin, chapter 8[*4], written over 2500 years ago. There is a tree that has been living since this the Discourses of Jin was written.

Forests bring awareness of our suffering society from a different perspective and show us the way to keep on living. As such, I believe forests can be said to be the superior doctor to make people think about "life" and "time".

Reference:
[*1] Yoshifumi Miyazaki and others "Analysis on the difference in the physiological effects of forest therapy among individuals" Nippon Eiseigaku Zassi, Japanese Journal of Hygiene, Vol.68, Supplement, March 2013
[*2] Medically, this may be called "the law of initial value" : Wilder（1931, 1967）
[*3] from "Xiaopin Fang" by Chen Yanzhi(Dr. Hiroshi Kosoto "Chugoku Igaku Koten to Nihon(Classic Chinese Medicine and Japan)" Hanawa Shobo publishing, 1996)
[*4] History book about China during the Spring and Autumn Period (-483BD), author is unknown but it is sometimes attributed to Zuo Qiuming of the state of Lu who wrote "The Zuo Zhohan: the Commentary of Zuo"

Written by Hideki Hirano 平野秀樹
Senior Fellow, The Tokyo Foundation
Vice Chief Director, Forest Therapy Society

Create Forests for Life
いのちの森をつくる

Akira Miyawaki 宮脇　昭

This is a wakeup call from Dr. Akira Miyawaki, the man who planted 40 million trees. He gives a warning for the future through an analysis of the state of forests.

What is a True Forest?
本物の森とはどういう森

A true forest is a native forest. People never tire of native-tree forests and their ecological potential. Native forests are long-lasting; they maintain their beauty for a long time. They possess a rich quality and can adapt to anything. They do not require any maintenance costs.

A native forest has native trees that create its potential, and therefore potential native forests can also be considered true forests. Vegetation suited to the local ecosystem is the foundation of native forests. For example, trees suited for Hokkaido are native Hokkaido trees and trees suited for Kyushu are native Kyushu trees.

Japanese society has gradually destroyed forests to build rice fields, farms, villages and towns instead. Fortunately, they have also maintained a tradition of creating local natural or at least quasi-natural forests by planting native trees well adapted to the land.

Japanese Tradition for Living in Harmony with Nature
自然を敬い、自然とともに生きてきた日本人

Native forests in Japan are referred to as *Chinju-no-mori*. The *jinja* inside the forests are considered sacred dwelling places of *kami* or departed souls. They keep out trespassers and maintain natural vegetation. *Chinju-no-mori* require no maintenance costs, they are hardy and long lasting, and resist invasion by foreign plants. In addition, *Chinju-no-mori* play a role in environmental conservation by offering sound protection, collecting dust, purifiying air, cultivating water sources as well as protecting nature from earthquakes, tsunamis, typhoons, gales, fires, and flooding. People feel safe with the forests and believe they are an integral part of their lives.

Many Japanese citizens perform ceremonial rituals and funerals in *Chinju-no-mori*. In times of happiness or sorrow, the forests have played a vital part in Japanese life. While it cannot be proven by science or technology, the close relationship and thousands of years of history between the Japanese people and forests have imprinted on the cultural DNA of Japanese society a deep bond that is difficult to deny.

Humanity is at Risk
今危険な状態の人類

About 4.6 billion years after the formation of earth and 3.9 to 4 billion years since the first life form came into existence, it can be argued that the ideal conditions in which living organisms satisfy each and every desire to the fullest are at risk. Living organisms mutually coexist by competing with each other or sacrificing themselves. Such is the healthiest symbiotic state.

Human beings are currently living in a precarious stage just like dinosaurs or mammoths faced before their extinction. In the 5 million years of human history, 4.99 million years were spent picking up acorns or young grasses and catching small fish in the stream. Now we are inundated with various materials and energy that satisfy the conditions humanity has strived for, but in the wake created a most dangerous ecological situation.

Let's look at parasitic worms as a comparative illustration. Parasitic worms suck the blood out of the host organism yet they do not suck enough blood to kill the host. If the host organism weakened, the worm would suck less blood and persevere to survive. The parasitic worms that continued to suck copious amounts of blood over the 4 million year history of living organisms, despite weakened hosts, eventually extinguished themselves . Such is also the delicate nature of survival between human beings and the forests they rely upon.

Living organisms live in a world where they compete, cohabitate and persevere. As a further example from nature, fallen seeds that wither or do not sprout will become natural fertilizer for other trees. Seeds that grow will do so slowly and will eventually become the sub-canopy in the forest. One day, the tall trees in the

Akira Miyawaki 宮脇 昭

Born in Okayama, Japan
Graduated Hiroshima University, Diploma in Biology
Botanist in plant ecology promoting planting all over the world
Current: Professor Emeritus of Yokohama National University,
    Director at IGES-Japanese Center for International Studies in Ecology (IGES-JISE)
Past: Visiting researcher, Federal Institute for Vegetation Mapping, Stolzenau, Germany
    Professor at Yokohama National University
    Director at Japanese Center for International Studies in Ecology (JISE)
Awards:
1970: Mainichi Publishing Award, 1991: Asahi Award, 1992: A Purple Ribbon Medal, 2000: Second Class Order of the Sacred Treasure, 2006: Blue Planet Prize
Publications:
"*Chinju-no-mori*" (2007, Shinchosha Publishing Co.,Ltd.) "Ki wo ueyo", Plant Trees (2006, Shinchosha Publishing Co.,Ltd.)"Sanbon no shokuju kara mori wa umareru" A Forest Begins with the Planting of Three Trees (2010, Shodensha Co., Ltd.)" Yonsenmanbon no ki wo ueta otoko ga nokosu kotoba" Words from the Man Who Planted 40 million Trees (2010, Kawade Shobo Shinsha)"Mori no Chikara" The Power of the Forest (2013, Kodansha Ltd.)" Gareki wo ikasu mori no bouhatei" Green Forest Wall built upon Debris (2012, Gakken Holdings Co.,Ltd.)

---

forest fall and the sub-canopy trees grow to replace the tall trees. This is the timeless teaching of the forest that humanity can learn from in order to prevent its own extinction.

Human Beings can Practice Tolerant Coexistence with The Forest
我々も少々我慢をして森と共存する

We know that in order to survive in this world, we need to practice tolerance for others who are different from us. We can survive by tolerating small inconveniences. Similarly, if you plant a tree, its leaves may fall or it may cast a shadow. Recognizing that as a small tolerable inconvenience instead of chopping the tree down permits ecological coexistence and a healthy way of life.

We will inevitably still require steel, cement and fossil fuels, and yet at the same time we must promote greenery and build multi-layered forests with native trees. The basis for human survival is that we continue living on this earth with limited land while balancing humanity with ecology.

The balancing act consists of the recycling system of production, consumption, decomposition, recomposition, and reducing the human footprint in the ecosystems of animals, vegetation and micro-organisms, both locally and globally. As the lead actors in the "drama" of life on Earth, we must not end this show as a tragedy. We should do our share by doing things that are easily done by everyone, like planting trees around us. If the tree eventually gets in the way, it can be cut down from the abundance we have created around it, but we must also use what we can from it and bury what we cannot use and plant a new tree in its place. Start by planting a sapling in an open space around you.

Native Forest Creation
森をくる

Creating a native forest is as natural as swimming in the ocean. It begins by choosing trees that are potential natural vegetation, which is defined as vegetation that would have been growing in the area without human intervention or management. In other words, it is important to select a species of tree that remains prevalent in the native forest. We should use several different types of trees to follow the natu-

Right photo; The tree planting at Iwanuma, Miyagi prefecture where hit by the tsunami in 2011.

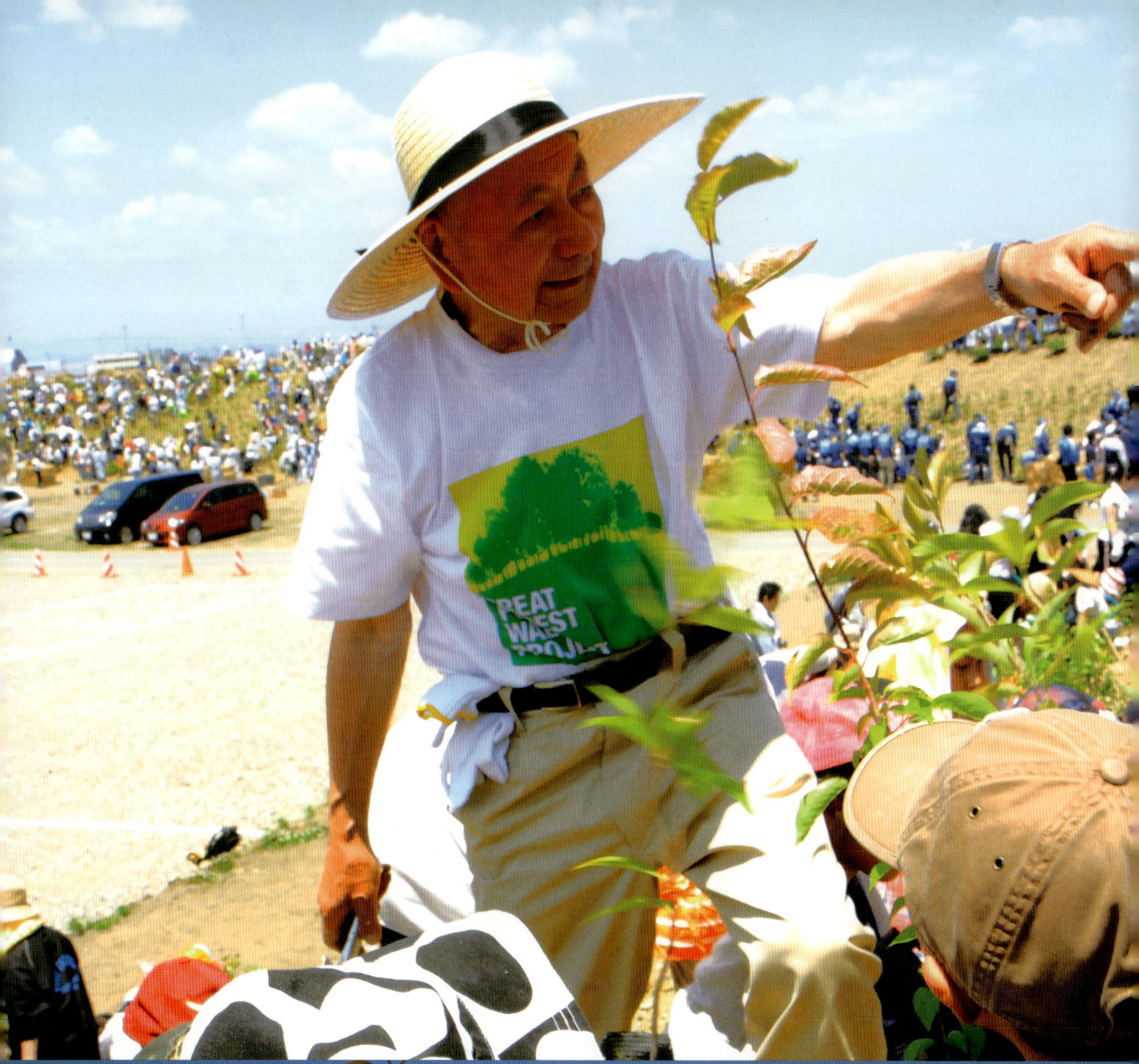

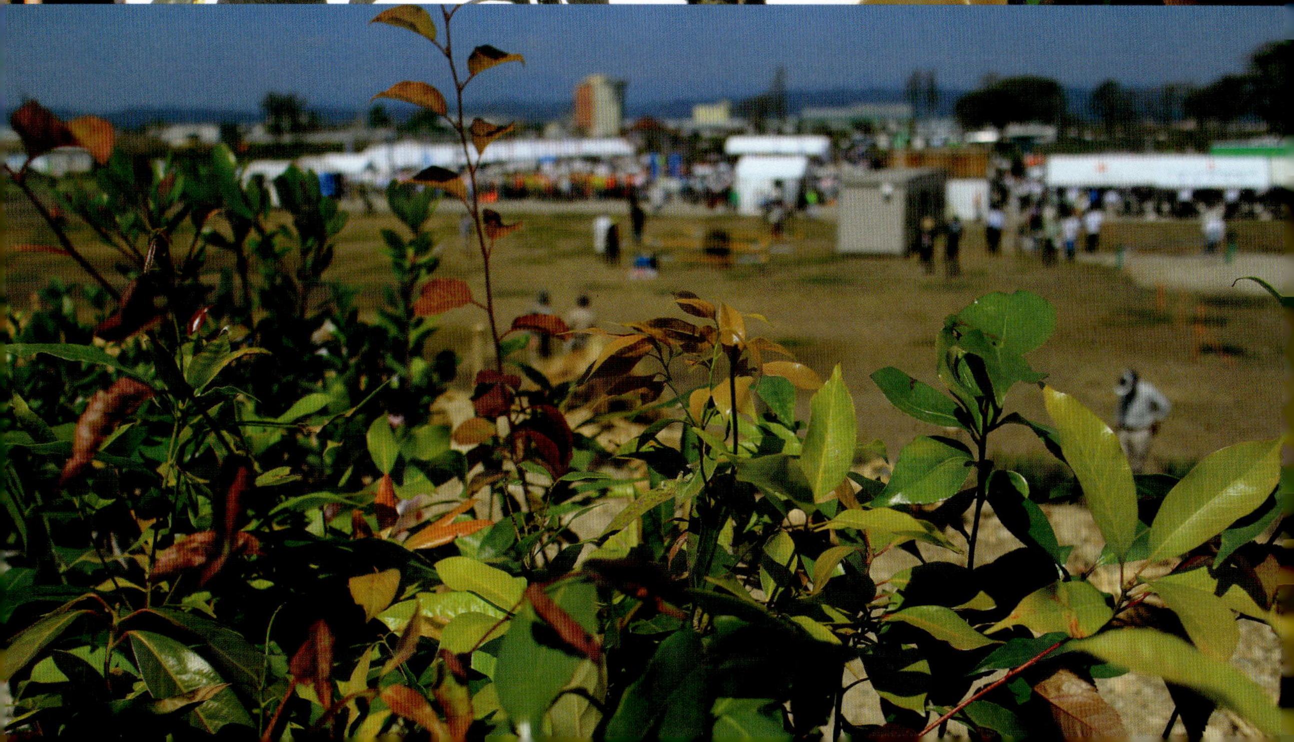

ral system of the forest, using mixed and close planting to aid in the prevention of disasters and the protection of the human population.

Without selecting tree species of the potential natural vegetation, forest restoration will not succeed. Large portions of the Japanese islands are covered by forest, but unfortunately, the original natural forests have been destroyed by human activity. Throughout history, forests have been uprooted for planting rice fields, yakihata (shifting cultivation), mining charcoal, and urbanization.

Unless you visit a site several times to become familiar with it, selecting tree species by latent natural vegetation could be akin to identifying the contents of your pocket without touching your clothes. *Chinju-no-mori* and forests that surround some old Japanese mansions have been very useful in providing keys for the study of potential natural vegetation.

Humans Cannot Survive Without The Forest
森がないと人類は生きていけない

Forests have 30 times more layers than the single-layer vegetation of a lawn, therefore forests protect human survival. Plants produce oxygen, clothing and furniture, as well as fossil fuels like carbonized pteridophyte ferns from prehistoric times. No animal on earth, including human beings, can survive without the forest.

When we destroy forests as if they are nuisances, cities decline and civilizations come to an end. In the historic Mediterranean region, since rainfall was very scarce, sclerophyll forests of cork oak trees once grew thickly. These cork oaks helped reduce transpiration with their hard and hairy leaves. This area was called the hardwood forest area. Human civilizations developed around these hardwood forest areas. Mesopotamia, Egypt, Greece, and the Roman Empire all saw their civilizations come to an end when they destroyed their hardwood forests. In fact, the ancient Egyptian pyramids were once surrounded by forests and greenery. It is believed that the forests were razed and pyramids were built on the land. It has been documented that during the Roman Empire there were evergreen trees in Egypt.

What We Can Do Now
我々が今できること

Compared to the planets' long life of 4.6 billion years, the time we live on earth is like a blink of an eye. We should therefore consider our lives in this context, including the intangible things we cannot measure. Such intangibles may be philosophy or religion.

What is happiness, an intangible might ask of us? That we have lived on this planet for so long is a miracle of the universe, and we are most fortunate because of it. Being alive after 4 billion years of continued evolution and DNA spinning without ever missing a link is a precarious gift, and therefore we must remember to live a life giving great care for the cycle of production, consumption, decomposition, recomposition, and reducing our human footprint in the ecosystems of all living matter. For me, in order to leave future generations our "life", "heart" and genes that have survived over 4 billion years, I plant trees of life.

Please do what you can right now for the future.

The Forest of Meiji Jingu
明治神宮の森

"Building" an eternal forest, chinju-no-mori
永遠に続く鎮守の森を「つくる」

The Forest of Meiji Jingu
明治神宮の森

The essay by Nobuko Nakamura　中村信子

Meiji Jingu is located in the center of the megalopolis of Tokyo and is surrounded by 700,000 square-meters of deep forest, called chinju-no-mori. Cross the bridge over the JR train of Harajuku station and you will reach an open area, which is the main entrance to Meiji Jingu. From this location, you can see a view of the Torii and the forest. The forest may seem as if it has been there for hundreds of years, treasured and revered by generations of people. However, the forest was man-made and planted by hand just 90 years ago.

Chinju-no-mori is a Native Forest of Native Trees
鎮守の森は天然自然の森

Japanese have believed that *Kami*s (sacred powers and spirits) reside in pure nature; therefore the people worship the forest, rivers, rocks, mountains and the sea as sacred places. As such, the sacred forest, *chinju-no-mori*, was created to enshrine *Kami*s when Meiji Jingu was built. As a sacred area, *chinju-no-mori* is to be free from human interference and people are not to enter to cut or plant trees. Therefore, it was necessary to create a forest that will regenerate naturally for years to come.

"If we choose vegetation that grows naturally in the environment of the region that will be the climax species (dominant species at the final stage of forest succession), those trees would seed by themselves and carry on to the next generations." said Koji Okisawa, forest keeper of Meiji Jingu.

For this kind of forest, native trees are planted that are adapted to the climate and soil type of the location. The land of this area is unique, with a dried volcanic ash layer called the Kanto loam layer. When the forest was being planted, a steam locomotive ran on the current JR line in front of Meiji Jingu, and a nearby thermal power station and tobacco factory created a smoky environment.

In botanical terms, this area is considered an evergreen broad-leaved forest region. It is also called an "oak area" because it is a climax forest of evergreen broad-leaved trees of which oak is the main species. Therefore the *chinju-no-mori* of Meiji Jingu is a different type of forest from the evergreen coniferous forests of Japanese cedar or Japanese cypress (*Hinoki* Cypress) in Jingu (Ise Jingu) or Nikko Toshogu.

Foundation of Meiji Jingu
明治神宮創建

In 1912, Emperor Meiji passed away and Fushimi-Momoyama of Kyoto was chosen

Right: Current south approach, Sndo.

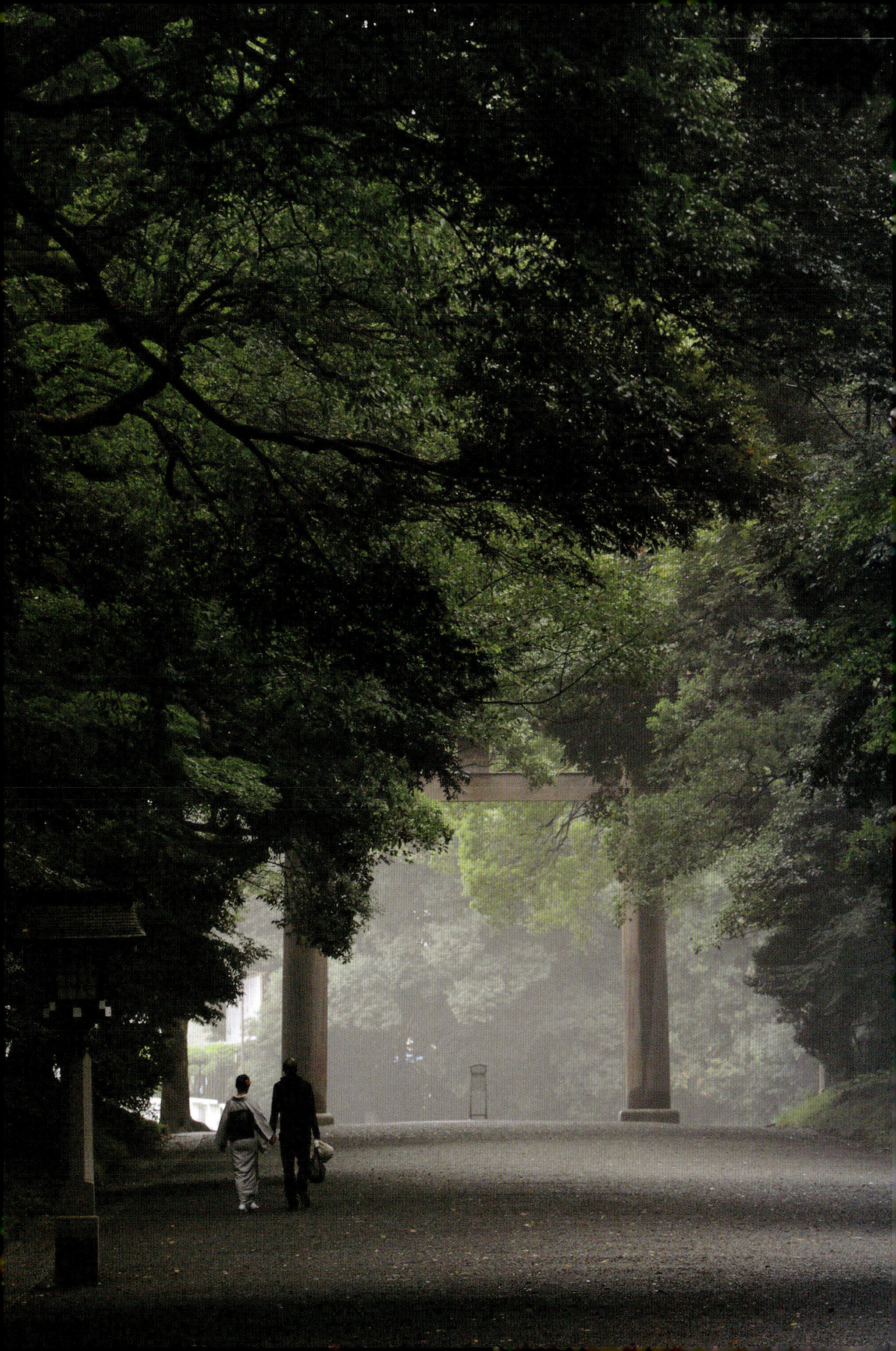

to be the location of his mausoleum. However, the voices of citizens who loved Emperor Meiji arose, calling for his mausoleum to be built in Tokyo. Because the location was already decided to be in Kyoto, the government decided to build a shrine to commemorate the late Emperor in Tokyo. The location was selected because it was part of the imperial estate and Emperor Meiji and Empress Shoken once lived in Tokyo and had close connections to the land.

The proposed site was large enough but was mostly wasteland and seedling fields. The forest comprised only one fifth to one sixth of the entire site.

The main members of the forest project were Dr. Seiroku Honda from Tokyo University, his assistant, Dr. Takanori Hongo, and their student, Keiji Uehara, who later came to be an honorary professor of Tokyo Agricultural University. They boldly took on the task of creating a forest. "We must create a forest in the devastated land!"

How did they design a forest?

The forest project members had visions of the forest in stages of 50, 100 and 150 years.

They advanced their idea to the final state of foresting and what kind of trees to plant by recording their ideas in a manual for anybody to read and learn from in the future. The book, titled "Meiji Jingu Forest Plan" (referred to as "Forest Plan" from here on) was beautifully bound in traditional Japanese style and published by Dr. Takanori Hongo in 1921, a year after the establishment of Meiji Jingu. This book shows the transition of the forest in four stages. (refer to picture 1)

BUILDING THE FOREST
森つくり

How was the forest built?

A total of 110,000 volunteers planted 100,000 donated trees through the National Tree Donation Project. Various trees chosen based on the Forest Plan were planted and the native trees on the site were kept as is. Deciduous trees such as zelkova, maple, and ginkgo were combined with evergreen broad-leafed trees. Deciduous tree leaves changed colors in autumn and were shed in winter. This allowed sunlight to reach the forest. The variety of trees allowed for diversity to create an

Part of "Meiji Jingu Forest Plan"

Left: Meiji Jingu ground before tree planting. The majority of the property was farmland.

picture 1

↑ Pine family

↑ Conifers except pines (Japanese cypress, sawara etc.)

● Evergreen broad-leaved trees (oak, castanopsises, cinnamomum camphora, etc.) and evergreen shrubs

First stage physiognomy
At the beginning of forest formation. Japanese red pines and Japanese black pines are the main trees planted. Conifers such as Japanese cypress, sawara, fir, and Japanese cedar make up a lower layer, and evergreen broad-leaved trees such as oak, castanopsises and cinnamomum camphora make up an even lower layer. Shrubbery and young evergreen tall trees grow in the lowest layer.

Second stage physiognomy
Pines, which are the main trees at the beginning of the forest, are expected to grow considerably. However, pines will eventually decline as they will be overtaken by the growth of Japanese cypress or sawara that come to dominate the crown canopy.

Third stage physiognomy
Within a hundred years of the establishment of the forest, oak, castanopsises and cinnamomum camphora become the dominant tree species and the forest is covered by evergreen broad-leaved trees. Japanese cedar, Japanese cypress, fir, a few Japanese black pines, and in certain places, old large trees of Japanese zelkova, muku (Aphananthe aspera) or gingko are mixed.

Fourth stage physiognomy
Decades to a hundred years after the third physiognomy, conifers will disappear. It will become a forest of oak, castanopsises and cinnamomum camphora. In particular, cinnamomum camphora will grow to be very large and dominant species, and will mix with naturally seeded young evergreen trees and small shrubs.

Above: The railway is the service line for construction.

Above: Japanese red pine transplanting

Left: Meiji Jingu ground after tree planting.

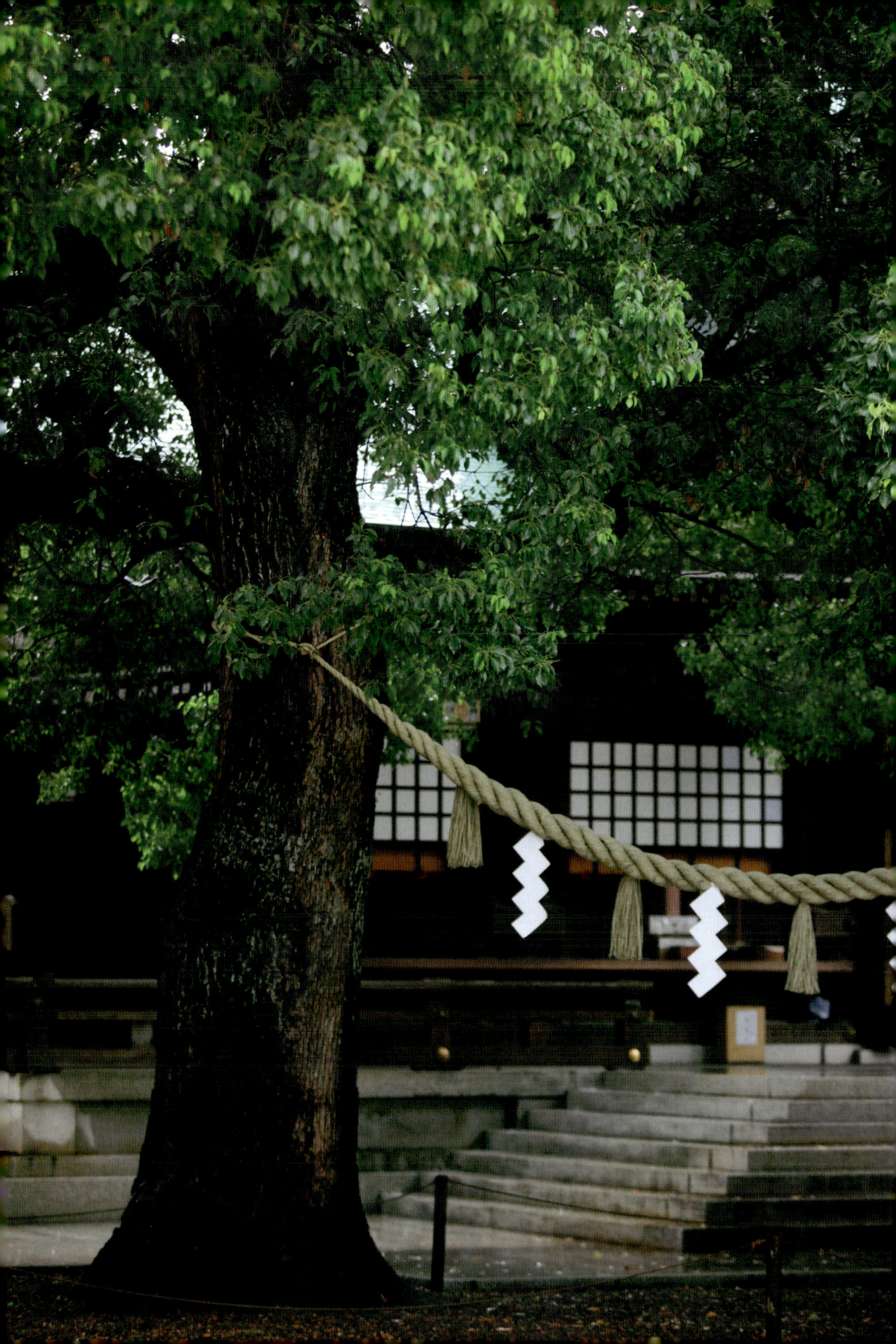

enchanting picture for visitors. Oaks were planted next to cinnamomum camphoras, zelkovas next to oaks, castanopsises next to zelkovas, and so on. Planting various trees grouped by species allows for easy extermination of insects.

"The Forest Plan is our bible. The more I read, the more I am impressed by the marvelous job done by people in the Meiji period (1868-1912)," said Koji Okisawa.

The Forest Plan covers possible issues and mitigation plans, including details such as where to tie up sacred horses in the event that they are used. Evidently those involved with this Forest Plan considered the future of the forest's creation. Interestingly enough, they seemed to know that evergreen broad-leaved trees were fire resistant. During World War II, Tokyo city was burned down by air attack but supposedly the forest of Meiji Jingu was mostly undamaged.

WACHING OVER THE FOREST
「見守る」ということ

Signs at the entrance of the Meiji Jingu forest warn "Do not break plants" and "Do not take any leaves or branches". The signs were placed in various places to help protect a natural life cycle: In the spring, plants sprout and undergo photosynthesis. Leaves fall during autumn and insects or bacteria in soil decompose fallen leaves to organic matter. Trees absorb them with water and continue to grow.

Visitors have faithfully followed the instructions of the signs and do not enter *chinju-no-mori*. Japanese have carried the tradition of not infringing upon *Kami*'s space. Aided by this sincere Japanese sense of respect, the forest has grown quicker than expected.

MAINTENANCE OF THE FOREST
森の「手入れ」

How is the forest maintained? The forest keeper gets involved only as needed and otherwise just watches over the forest, allowing it to maintain itself. For example, if any branch or trunk of a tree gets in the way of the sidewalk or roadway, the keeper simply picks it up and returns it to the forest. The Meiji Jingu precincts are swept every day. The leaf and branch debris are returned to the forest, as this is a part of protecting the forest cycle.

Dead and removed trees are used one last time before being returned to the soil, taking the reusable parts of the trees to create signs, fences, or for wooden foundations for pathways. Better quality wood is used to make wooden bells called "Kodama". Such is the method of Meiji Jingu forest's maintenance so as not to bother the forest's natural cycle.

FOREST KEEPER KOJI OKISAWA'S THOUGHTS
森の番人をしている沖沢幸二氏の思い

Interviewing Koji Okisawa, the forest keeper of Meiji Jingu, I understood how much he devotes himself to and respects the forest, preserving trees with great care. His story of the forest warmed my heart.

Left: In front of main building.

"Instead of saying things such as a tree has 'died' or 'fallen', I prefer to say that the tree has 'fulfilled its duties'. When cutting it down, I say to the tree, 'Good work,' or 'thank you for the hard work'. The amount of time we can interact with a tree is just a tiny fraction of its several-centuries-long life span. Within our short time, it is our job to watch over the forest so it carries on to its next generation and also carries on our concepts and methods of maintenance to our next generation.

Every page in the Forest Plan is sensible and accurate. It is really amazing how marvelous of a book our predecessors left us. Anybody whose work is related to the forest should keep this book beside them, reading it over and can refer to it while taking care of the forest. Over time, the role of forest-keeper will be passed on from person to person, but as long as they read this book and follow its philosophy, the forest can continuously be maintained with the same attitude.

The forest of Meiji Jingu is a special place for people who work with forests. Our predecessors learned about the forest from those in emperors' mausoleums or old *jinja* that remained untouched for generations. Using that knowledge as a foundation, they worked hard to create a self-sustaining forest and they succeeded. The fact that the forest is progressing as they initially planned is an admirable thing.

Ninety years have passed since the forest was made. It is nearing the state of an ideal forest at a quicker rate than anticipated. One of the original project members, Uehara, had seen the forest's progress and remarked that the entire forest plan may be accomplished in 100 years, instead of the originally expected 150 years. Since the forest has reached the state of the fourth phase, there are some that say the forest is completed. But we think we are only at the starting line of *chinju-no-mori*. The forest will raise its next generation and various species of trees will fluctuate as they survive competitively. I believe that in 500-600 years, the forest will become a spacious forest, fitting of the name, *chinju-no-mori*."

So, ninety years after the forest was planted in the middle of a metropolitan area, the forest may still considered to be young. Although it is still young, walking along its treasured pathway conjures a warm and pure feeling. By aging another 100 and 200 years from now, the forest will become even more powerful. In order for this to happen, we must pass on the same attitude towards to the forest to future generations.

Left: All of the fallen leaves are swept and returned to the forest. They are called "Mr. sweepers".

Right: Meiji Jingu Gyoen, at the beginning of June when the irises start to blossom.

Summary of Forests in Japan

Total Land Area   37,280,000 ha   100%
Forest Area   25,090,000 ha   67%
Planted Forest   10,346,000 ha   27%

National Forestry
Japan has 7,580,000 ha of designated nation-owned forest. This makes up 20% of the total land area or 30% of the overall forest area. Most of the forests play a role in national land preservation, headwater conservation, and natural environmental preservation.

1. Forest Reserve 6,830,000 ha

    A Forest Reserve is an area of land that is designated by a minister of Agriculture, Forestry and Fisheries or a governor for its specific role in headwater conservation, disaster protection, and environmental protection.

2. World Natural Heritage Forest

    Shirakami-Sanchi, Yakushima, Shiretoko, Ogasawara Islands etc.
    These natural forests are part of the national forest inventory and registered as World Natural Heritage sites. Rare wild animals live in these forests.

3. UNESCO Park (Biosphere Reserve)

    Biosphere Reserves are areas of terrestrial and coastal ecosystems promoting solutions to reconcile the conservation of biodiversity with their sustainable use.
    Areas that serve the functions below are registered as biosphere reserves.

    conservation of biodiversity
    social and economic development
    academic research support

    UNESCO biosphere reserve park sites in Japan
    Shiga Highland (Nagano and Gunma prefecture)
    Mount Hakuasan (Gifu, Ishikawa, Toyama, Fukui Prefectue)
    Mount Odaihara & Mount Omine (Nara, Mie prefecture)
    Yakushima (Kagohima prefecture) registered in 1980 (Showa 55).
    Aya (Aya city, Miyagi prefecture) registered in 2012 (Heisei 24).

Planted Forest
As of 2014, the Japan Forestry Agency has reported that since World War II, 10,346,000 ha of forest have been planted to supply lumber in Japan. These forests cover 27% of the entire land area of Japan. It has been over 50 to 60 years since they were planted, and they are now considered ready for logging for lumber. However, the availability of cheap imported lumber has created a decrease in the price of domestic lumber. Additionally, labor and material cost increases and profit decreases have caused the forest industry in Japan to slow down, resulting in an overgrowth of cedar groves throughout the nation. Currently, the self-sufficiency rate of lumber in Japan is about 28% due to the increase of foreign lumber imports. The Forestry Agency targets its self-sufficiency rate to increase to 50% by the year 2020 (Heisei 32).
The reformation plan is as follows:

    1.Preparation of a system that reliably carries out an appropriate forestry enforcement.
    2.Preparation of conditions where the probability of a low-cost work system can be established in a wide area.
    3.Nurturing human resources that will shoulder the forestry industry.
    4.Creation of a system where domestic lumber can be efficiently processed and distributed, in addition to an increase in demand for lumber usage, etc.

Figures taken from the Japan Forestry Agency statistics report, March 31, 2007 (Heisei 19)

Glossary

CHINJU-NO-MORI	Sacred forest: Refer to page 3.
GOHEI	Ritual wand.
GUJI	Chief priest of jinja.
HAKAMA	Loose trousers forming part of Japanese formal dress.
HAIDEN	A building where people worship.
HIMOROGI	A temporarily erected tree branch to which summons the kami.
HOKORA	A miniature jinja structure.
HONDEN	The central building where kami dwell.
HONGU	Main jinja building.
HORYUJI	The world's oldest wooden architectural building located in Nara Prefecture built in 607 (15th year of the Emperor Suiko's Era).
IWAKURA	Rocks which summon the kami.
JINJA	Places where the kami reside: Refer to page 4.
KAGURA	Ancient dance and music dedicated to kami.
KAMI	Sacred powers and spirits: Refer to page 3.
KAYABUKIYANE	A method of making rooftops, made from Japanese nutmeg tree, silver, grass, or cogon grass. It is considered to be a primitive roof style.
KITO-DEN	Prayer hall.
MANYOSHU	The oldest collection of Japanese poems edited in the middle of the 8th century.
MATSURI	Sacred rituals with religious service and entertainment for kami.
MIKOSHI	A portable jinja where kami are carried in a procession during festivals.
MIKOSHI-TOGYO	Mikoshi parade.
OBON	An event to commemorate and honor the spirits of ancestors.
OFUDA	Talisman that acts as a symbol of a kami to protect a household.
OHIGAN	A holiday observed during the Spring and Fall Equinoxes.
OKUMIYA	Innermost jinja.
OMAMORI	Protective amulet or charm.
OSAISEN	Offerings.
RYUJIN	Dragon kami.
SAKAKI	Sacred evergreen tree, from which usually branches are used.
SANDO	A path built for jinja visitors, which leads from the entrance to the jinja buildings.
SATOYAMA	Woodlands located between residential area and the undeveloped nature.
SHIMENAWA	Braided rice straw rope placed around the objects.
SHINSHOKU	Priests.
TEMIZUYA	Fountain at the entrance to jinja, through which one must first be purified.
TORII	A gate for entrance of jinja.
UJIGAMI	Local kami.
UJIKO	Parishioners of a jinja.
YOIMATURI	A small matsuri on the eve of the matsuri.
YUKATA	A casual summer kimono.

VOICE OF THE FOREST 神の森

2014年4月20日　初版発行

発売／株式会社ヴォイス 〒106-0031 東京都港区西麻布 3-24-17 広瀬ビル 2F
　　　　　　　　　TEL:03-3408-7473　FAX:03-5411-1939
　　　　　　　　　http://www.voice-inc.co.jp/

印刷・製本／株式会社光邦

ISBN 978-4-89976-419-9
禁無断転載・複製

発 行 人	大森浩司
編　　集	中村信子
編集協力	有限会社 United Works　http://www.unitedworks.net
英文編集	カリ・クセラ、スーザン・ハムレ
翻　　訳	タミー・フォン
写　　真	小野祐次（表紙、6～13、16～17、59～61、85～87、90） 松尾成美（32～39）、中村信子、フォトストック
イラスト	稲澤美穂子
協　　力	神宮司廳、神社本廳、明治神宮、鶴岡八幡宮、秩父神社、 貴船神社、下鴨神社、富士山本宮浅間大社、白山比咩神社、 彌彦神社、竃門神社、八重垣神社、スティーブ・ストリックランド、 メーガン・ミッチェル　他　順不同

VOICE OF THE FOREST

First edition, 20 April, 2014

Published by VOICE Inc.
Hirose Bldg.2F,3-24-17 Nishi-Azabu Minato-ku, Tokyo 106-0031 Japan
TEL:+81-3-3408-7473　FAX:+81-3-5411-1939　http://www.voice-inc.co.jp/

Printed in Japan

ISBN 978-4-89976-419-9
2014 by VOICE Inc.; all right reserved.

Publisher	Koji Omori
Editor in chief	Nobuko Nakamura
	United Works Corp.　http://www.unitedworks.net
English editing	Kali Kucera, Susan Hamre
Translation	Tammy Fung
Photography	Yuji Ono（Cover,6-13,16-17,59-61,85-87,90) Narumi Matsuo（32-39),Nobuko Nakamura and library
Illustration	Mihoko Inazawa

In corporation with Jingu Administration Office, Jinja Honcho, Meiji Jingu, Tsurugaoka Hachimangu, Chichibu Jinja, Kifune Jinja, Shimogamo Jinja, Fujisan Hongu Sengen Taisha, Shirayamahime Jinja, Yahiko Jinja, Kamado Jinja, Yaegaki Jinja, Steve Strickland, Megan Mitchell and others